ENTOMOLOGICAL SOCIETY OF AMERICA

Handbook of Small Grain Insects

EDITED BY

G. David Buntin
Keith S. Pike
Michael J. Weiss
James A. Webster

Entomological Society of America
10001 Derekwood Lane, Suite 100
Lanham, MD 20706-4876 USA
Phone: 301-731-4535
Fax: 301-731-4538
www.entsoc.org

The Entomological Society of America (ESA) is a not-for-profit organization serving the scientific and professional needs of entomologists and individuals in related disciplines throughout the world. Formed in 1953 by the consolidation of the American Association of Economic Entomologists (founded in 1898) and the former Entomological Society of America (founded in 1906), ESA is the largest international association of entomologists. It has approximately 6,000 members at colleges and universities, state and federal departments of agriculture, other government agencies, health agencies, private industry, parks, and museums.

This book was published in cooperation with The American Phytopathological Society (APS). APS is an international scientific organization devoted to the study of plant diseases and their control. The Society was founded in 1908 and has grown from 130 charter members to nearly 5,000 plant pathologists and scientists worldwide. APS provides information on the latest developments and research advances in plant health science through its journals and its publishing arm, APS PRESS.

The American Phytopathological Society
3340 Pilot Knob Road
St. Paul, MN 55121 USA
Phone: 651-454-7250
Fax: 651-454-0766
www.apsnet.org

To Order This Book
1-800-328-7560 or www.shopapspress.org
ESA and APS members are entitled to a 10% discount and must reference their membership status at the time the order is placed.

Front cover photograph: Wheat stem sawfly, *Cephus cinctus,* by Jim Kalisch, University of Nebraska.

COPYRIGHT © 2007 BY THE ENTOMOLOGICAL SOCIETY OF AMERICA
10001 DEREKWOOD LANE, SUITE 100, LANHAM, MD 20706, USA
ALL RIGHTS RESERVED

PRINTED AND BOUND IN CHINA

ISBN 9780977620913

LIBRARY OF CONGRESS CONTROL NUMBER: 2007920120

Contents

Acknowledgments	iv
About the Editors/Contributing Authors	v
How to Use this Book	1
An Introduction to Small Grains	2
Importance and Types of Small Grains	2
Growth and Development of Small Grains	4
Small Grain Production Practices	7
Insect and Mite Injury to Small Grains	12
Small Grain Pest Management	14
Principles of Small Grain Pest Management	14
Sampling and Decision Making in Arthropod Pest Management	14
Pest Management Tactics for Small Grain Arthropod Pests	16
Diseases and Arthropod Pest Management	20
Weeds and Arthropod Pest Management	22
Identification of Arthropods and Diagnosis of Injury	24
Small Grain Arthropod Pests by Injury Type	25
Small Grain Arthropod Pests by Scientific Classification	26
Key to Insect Injury to Wheat	27
Key to Insect and Mite Pests of Small Grains	30
Pest Information	37
Introduction	37
Aphids	37
Bird Cherry-Oat Aphid	37
Corn Leaf Aphid	38
English Grain Aphid	39
Greenbug	40
Russian Wheat Aphid	42
Yellow Sugarcane Aphid	43
Miscellaneous Aphids	44
Armyworms	46
Armyworm	46
Fall Armyworm	46
Wheat Head Armyworm	47
Yellowstriped Armyworm	48
Billbugs	49
Blister Beetles	49
Cereal Leaf Beetle	50
Chinch Bug	52
Cutworms	53
Army Cutworm	53
Pale Western Cutworm	54
Flea Beetles	55
Frit Fly	56
Grasshoppers	56
Hessian Fly	58
Leaf Sawflies	62
Leafhoppers and Planthoppers	63
Leafminers/Grass Sheathminer	63
Lesser Cornstalk Borer	64
Mormon Cricket	65
Plant Bugs	66
Seedcorn Maggot	67
Stalk-Boring Caterpillars	68
European and Native Corn Borer	68
Stalk Borer	69
Stink Bugs	70
Thrips	72
Wheat Jointworm	73
Wheat Midge	73
Wheat Mites	75
Wheat Curl Mite	75
Brown Wheat Mite	77
Banks Grass Mite	78
Winter Grain Mite	79
Wheat Stem Maggot	79
Wheat Stem Sawfly	80
Wheat Strawworm	82
White Grubs	83
Wireworms and False Wireworms	84
Wireworms	84
False Wireworms	85
Insect Pests Outside of North America	86
Sunn Pest and Cereal Bugs	86
Wheat Ground Beetle	87
Ground Pearls	88
Barley Stem Gall Midge	89
Barley Shoot Fly	90
Black Fly	90
Migratory Locust	91
Beneficial Organisms	93
Entomopathogens	93
Parasitoids	94
Predators	96
References	101
Glossary	107
Sources of Information	112
Index	117

Acknowledgments

Many individuals were involved in the development of the *Handbook of Small Grain Insects*. These people include the 54 authors who contributed sections or chapters. Their expertise, hard work, patience, and supportive attitude are greatly appreciated. We thank Leon Higley (University of Nebraska) for developing the maps used in this handbook. We also thank the authors and other people who contributed photographs for use in the handbook. This handbook would not be possible without their contribution.

In particular, we thank Wendell Morrill (Montana State University) for his contributions as an author and for providing photos of numerous insects and their damage. We also thank Jim Kalisch (University of Nebraska), Marlin Rice (Iowa State University), Jerome Grant (University of Tennessee), V. H. Beregovoy (Oklahoma State University), and Scott Bauer (USDA-ARS) for providing photos of arthropods.

We wish to thank former ESA president Lowell Nault, Ohio State University, for his vision and efforts to initiate the handbook series. We also thank the various members who have served on the ESA Handbook Editorial Committee for their guidance and assistance. We also thank editors of previous handbooks including Leon Higley, David Boethel, (Louisiana State University), Michael Gray (University of Illinois), Kevin Steffey (University of Illinois), and Marlin Rice (Iowa State University) for their assistance, input, and encouragement during the development of our handbook.

In addition, we thank staff of the Entomological Society of America for their hard work and assistance during the production of this handbook. In particular, we thank Alan Kahan for his expertise, help and support during the completion of the handbook.

This handbook has been in development for a number of years. We sincerely appreciate the patience of the authors and ESA membership as we assembled and completed this handbook. We hope the wait was worthwhile and that this handbook is a useful and valuable addition to the series.

The Editors
January 2007

About the Editors

David Buntin is a professor in the Department of Entomology at the University of Georgia at Griffin. He received his B.S. in Entomology and Applied Ecology from the University of Delaware (1977) and an M.S. and Ph.D. in Entomology from Iowa State University (1980 and 1984). Since 1984, Dr. Buntin has held a research and extension appointment at the University of Georgia. His program focuses on insect management in grain crops including wheat and other cereal grains, corn, sorghum, and canola. He also recently has begun teaching IPM at the UGA-Griffin campus.

Keith Pike is a professor in the Department of Entomology at Washington State University at Prosser. He received his B.S. in Entomology from Utah State University (1971), and an M.S. and Ph.D. in Entomology from the University of Wyoming (1973 and 1974). He served as a research associate with the University of Nebraska (1974-1975), and since 1976, has held a research appointment at Washington State University. His program focuses on insect pest management in small grains (wheat and barley) and selected irrigated crops with an emphasis on aphids and aphid parasitic biocontrol.

Mike Weiss was a faculty member at the Eastern Agricultural Research Center and North Dakota State University where he had responsibility for developing pest management strategies for small grain insects, primarily for spring wheat and barley. He has served on the faculty at the University of Idaho and presently is a member of Auburn University faculty. He has been a member of the Entomological Society of America since 1977 and has served in various leadership roles at the national and regional level.

James (Jim) Webster is a retired USDA-ARS Research Entomologist and an Adjunct Professor of Entomology at Oklahoma State University. He received his B.S. and M.S. from the University of Kentucky (Zoology 1961, Entomology 1964) and Ph.D. in Entomology from Kansas State University (1968). Dr. Webster conducted plant resistance research during his 32-year career with USDA-ARS. His research focused on cereal leaf beetles in East Lansing, MI (1968-1981) and cereal aphids in Stillwater, OK (1981-2000). He also served as Laboratory Director of the USDA-ARS Plant Science and Water Conservation Research Laboratory and Research Leader of the Plant Science Unit in Stillwater, OK (1993-1999).

Contributing Authors

Sue L. Blodgett
Plant Science Department
South Dakota State University
Brookings, SD 57007

Mustapha El Bouhssini
Germplasm Program
ICARDA
P.O. Box 5466
Aleppo, Syria

H. Leroy Brooks
Department of Entomology
Kansas State University
Manhattan, KS 66506

Michael D. Bryan
USDA–APHIS
Niles Plant Protection Center
2534 South 11th St.
Niles, MI 49120

G. David Buntin
Department of Entomology
Georgia Experiment Station
University of Georgia
Griffin, GA 30223

John D. Burd
USDA–ARS
Plant Science & Water Conservation Research Laboratory
1301 N. Western St.
Stillwater, OK 74075

Jay W. Chapin
Clemson University
Edisto Research and Education Center
64 Research Road
Blackville, SC 29817

Barry M. Cunfer (retired)
Department of Plant Pathology
University of Georgia
Griffin Campus
Griffin, GA 30223

Norman C. Elliott
USDA–ARS, SPA, Plant Science Research Laboratory
1301 N. Western St.
Stillwater, OK 74075

Robert H. Elliott
Agriculture and Agri-Food Canada
Saskatoon Research Centre
107 Science Place
Saskatoon, SK Canada S7N 0X2

Kathy L. Flanders
Auburn University
Department of Entomology and Plant Pathology
208A Extension Hall
Auburn, AL 36849-5629

B. Wade French
USDA–ARS, SPA
Plant Science Research Laboratory
1301 N. Western St.
Stillwater, OK 74075

Joseph E. Funderburk
North Florida Research & Education Center
University of Florida
155 Research Road
Quincy, FL 32351

Kristopher L. Giles
Department of Entomology
Oklahoma State University
127 Noble Research Center
Stillwater, OK 74078-3033

Phillip A. Glogoza
University of Minnesota Extension Service
715 11th St. N., Suite 107C
Moorhead, MN 56560

Jeremy K. Greene
Clemson University
Edisto Research and Education Center
64 Research Road
Blackville, SC 29817

Matthew H. Greenstone
USDA–ARS, PSI
Insect Biocontrol Lab
10300 Baltimore Ave.
Beltsville, MD 20705

Contributing Authors

Susan E. Halbert
Florida Department of Agriculture and Consumer Services
Division of Plant Industry
P.O. Box 147100
Gainesville, FL 32614-7100

Gary L. Hein
University of Nebraska
Panhandle Research & Extension Center
4502 Ave. I
Scottsbluff, NE 69361

Richard L. Hellmich
USDA–ARS
Corn Insects and Crop Genetics Research Unit
110 Genetics Lab, Insectary Building
Iowa State University
Ames, IA 50011

D. Ames Herbert, Jr.
Virginia Tech
Tidewater Agricultural Research and Extension Center
6321 Holland Rd.
Suffolk, VA 23437

Donald R. Johnson
45 Woody Lane
Cabot, AR 72023

Gregory D. Johnson
Montana State University
Department of Rangeland Science
Bozeman, MT 59717

James B. Johnson
University of Idaho
Division of Entomology
Moscow, ID 83844-2339

Jerry W. Johnson
University of Georgia
Department of Crop and Soil Sciences
Griffin Campus
Griffin, GA 30223

Jay B. Karren
Utah State University
Department of Biology
Logan, UT 84322-5305

S. Dean Kindler
USDA–ARS
1301 N. Western St.
Stillwater, OK 74075

Eugene G. Krenzer, Jr.
Oklahoma State University
Department of Plant and Soil Sciences
Stillwater, OK 74078

Saadia Lahloui
INRA (Institut National de Récherches Agronomique)
Centre Aridoculture
B.P. 589
Settat, Morocco

Robert J. Lamb
Agriculture and Agri-Food Canada
Cereal Research Centre
195 Dafoe Rd.
Winnipeg, MN Canada R3T 2M9

John D. Lattin
Oregon State University
Systematic Entomology Laboratory
Department of Entomology
Corvallis OR 97331-2907

Z B Mayo
University of Nebraska
Agricultural Research Division
207 Agricultural Hall
Lincoln, NE 68583-0704

Gerald J. Michels, Jr.
Texas A&M University
Texas Agricultural Experiment Station
2301 Experiment Station Rd.
Bushland, TX 79012

Ross H. Miller
University of Guam
College of Agriculture and Life Sciences
Agricultural Experiment Station
Mangilao, Guam 96923

Wendell L. Morrill
Montana State University
Land Resources and Environmental Science
Bozeman, MT 59717

Owen Olfert
Agriculture and Agri-Food Canada
Saskatoon Research Centre
107 Science Place
Saskatoon, SK Canada S7N 0X2

Frank B. Peairs
Colorado State University
Department of Bioagricultural Sciences and Pest Management
Fort Collins, CO 80523-1177

Contributing Authors

Keith S. Pike
Washington State University
Irrigated Agriculture & Extension Center
24106 N. Bunn Rd.
Prosser, WA 99350

Tadeusz J. Poprawski
USDA–ARS Subtropical Agricultural Research Center
Beneficial Insects Research Unit
2413 East Highway 83
Weslaco, TX 78596

David R. Porter
USDA–ARS
1301 N. Western Road
Stillwater, OK 74075

Roger H. Ratcliffe
3500 Canterbury Drive
Lafayette, IN 47909

Larry Robertson
Department of Plant, Soil, and Entomological Sciences
University of Idaho
Moscow, ID 83844

Tom A. Royer
Oklahoma State University
Department of Entomology
127 Noble Research Center
Stillwater, OK 74078-3033

Larry Sandvol
University of Idaho Research & Extension Center
P.O. Box AA
Aberdeen, ID 83210

Phillip E. Sloderbeck
Kansas State University
4500 E. Mary Street
Garden City, KS 67846

C. Michael Smith
Kansas State University
Department of Entomology
123 Waters Hall
Manhattan, KS 66506-4004

David R. Smith
National Museum of Natural History
Systematic Entomology Laboratory
Smithsonian Institution MRC-168
Washington, DC 20560

Petr Starý
Czech Academy of Sciences
Institute of Entomology
Ceské Budejovice, Czech Republic

Glenn Studebaker
University of Arkansas
PO Box 48
1241 West County Road 780
Keiser, AR 72351

William Turner
Washington State University
Department of Entomology
Pullman, WA 99164-6382

John W. Van Duyn
North Carolina State University
Vernon G. James Research and Extension Center
207 Research Station Road
Plymouth, NC 27962

Michael J. Weiss
Golden Harvest Seeds
502 E. Broadway #1
Decorah, IA 52101

Gerald E. Wilde
Kansas State University
Department of Entomology
123 Waters Hall
Manhattan, KS 66506

Stephen P. Wraight
USDA–ARS Plant Soil and Nutrition Laboratory
Plant Protection Research Unit
Tower Road
Ithaca, NY 14850

How to Use This Handbook

By G. David Buntin

This handbook is designed primarily for the practitioners of integrated pest management (IPM) programs in small grains, growers, crop consultants, extension agents, and company agronomists and sales representatives. Students and researchers also will find valuable information in this handbook. Our primary objective is to provide fundamental and useful information about managing small grain insects throughout the United States and Canada. Although this handbook focuses on insect pests of small grains, noninsect pests (e.g., mites) also are covered. Crops covered in this handbook are wheat, barley, oats, rye, and triticale with an emphasis on wheat. Rice, millet and other grain crops are not covered.

The first three sections of the handbook provide information about small grains and their production, principles and practices of small grain insect management, and an overview of the pest injury to small grains by insects, weeds, and plant pathogens. The remainder of the handbook is devoted to discussions of insect and mite pests of small grains and to beneficial organisms. Authors are identified at the end of each section, and a list of references for additional information is provided for most sections. A glossary provides definitions for unfamiliar words. A section is provided on sources of local information throughout the United States and Canada.

The introduction to identification of insects and diagnosis of injury explains how to use the list of small grain insect pests grouped by injury classification, the key to small grain injury caused by insect pests, and the key to small grain pests. The list and keys provide directions to sections in the handbook in which color photographs of the insects and examples of injury to small grains accompany detailed discussions about the pests. The keys and color photographs provide assistance for identifying insects and diagnosing injury caused by insects in small grains.

Insect pests that attack small grains are discussed in sections arranged alphabetically by common name for easy reference. Within each section, information is provided on scientific classification, origin and distribution, description (aids in identification), pest status (relative importance of the pest), injury (aids in diagnosis), life history, and management. Information on important natural enemies also may be included. We have attempted to include all known insect and mite pests that are likely to be encountered in small grains in the United States and Canada. However, incidental insects found in small grains and possibly some pests with a limited distribution may not be included.

Suggestions for management are included as an overview of control tactics that are effective for a specific pest. If insecticides are useful for managing a pest, they may be mentioned; but trade names are not identified. Economic thresholds, when known, also are provided. However, economic thresholds and management recommendations may vary from one region of North America to another. Refer to a state's Cooperative Extension Service guidelines for managing insects in small grains and to get information about regionally accepted economic thresholds and specific insecticide recommendations.

The section on beneficial organisms, including insect pathogens (entomopathogens), parasitoids, and predators, provides information about their potential for natural or applied biological control of small grain insect pests. The information varies among organisms because of the diversity of their impact in small grain cropping systems.

An Introduction to Small Grains

Importance and Types of Small Grains

Small grains are big business for agriculture in North America. Small grain products form the foundation of the daily nutritional sustenance for most of humankind. The term "small grains" refers to the small-seeded cereals, which include wheat, barley, oats, rye, and triticale. World and North American production figures for all grain crops, including corn and rice, are listed in Table 1.

Wheat

Wheat, *Triticum aestivum* L., is a genetically diverse and ancient cereal crop. It evolved from wild grasses in an area near the Tigris–Euphrates River basin, and was first used by humans perhaps as early as 15,000 B.C. Wheat was brought to North America from Europe by colonists settling in the eastern United States during the 17th century. Around 1874, the Mennonites settling in Kansas brought with them a hard red winter wheat cultivar from the Ukraine named 'Turkey'. The introduction of this cultivar ushered in the expansion of hard red winter wheat production throughout the Great Plains, prompting the reference to this region as the "Bread Basket of the World."

Globally, wheat provides more nourishment than any other food source. In the United States, 2003 per capita consumption of wheat flour and cereal was 142 lb, making wheat the most consumed cereal grain. The value of U.S. wheat production for 2003 was $7.9 billion (Table 2). More than 2 billion bushels were harvested from ~53 million acres nationwide. Wheat is the second most valuable crop in the United States behind corn, but it is the leading cereal crop for export. Almost 23 million metric tons of wheat and flour worth an estimated $3.9 billion were exported in 2003.

Wheat can be classified by its growing season (winter or spring), seed hardness (hard or soft), and seed color (red or white). Winter wheat is planted in the fall. In the Southern Plains, it is typically grazed by cattle and then harvested in late spring or summer. Spring wheat is planted in spring and harvested in late summer or early fall. There are six basic classes of wheat: hard red winter, hard red spring, soft red winter, durum, hard white, and soft white. Production areas and uses of each market class are presented in Fig. 1. About 95% of all Canadian wheat is hard red spring and is produced in three prairie provinces (Manitoba, Saskatchewan, and Alberta).

Barley

Barley, *Hordeum vulgare* L., is thought to have been domesticated about 7,000 B.C., but it was not used for brewing alcoholic beverages until about 6,000 B.C. It is the world's fourth most important cereal after corn, wheat, and rice. Canada was the largest producing country in North America, with 566 million bushels in 2003. In the United States, the 2003 value of barley production was about $755 million, making it the fifth most important cereal (Table 2). These figures do not take into account the

Table 1. Worldwide and North American production of major grain crops in 2003.

Grain crop	World production (billion bushels)	North America production (million bushels)
Corn	24.4	11,347
Wheat	20.3	3,316
Rice	19.0	323
Barley	6.4	885
Oats	1.8	405
Rye	0.6	20

Table 2. Production and value of major grain crops in the United States in 2003.

Grain crop	Grain test weight (lb/bushel)	Production (million bushels)	Value of production (million $ U.S.)
Corn	56	10,089	24,477
Wheat	60	2,345	7,929
Rice	45	313	1,629
Sorghum	56	411	965
Barley	48	278	755
Oats	32	144	225
Rye	56	9	25

An Introduction to Small Grains

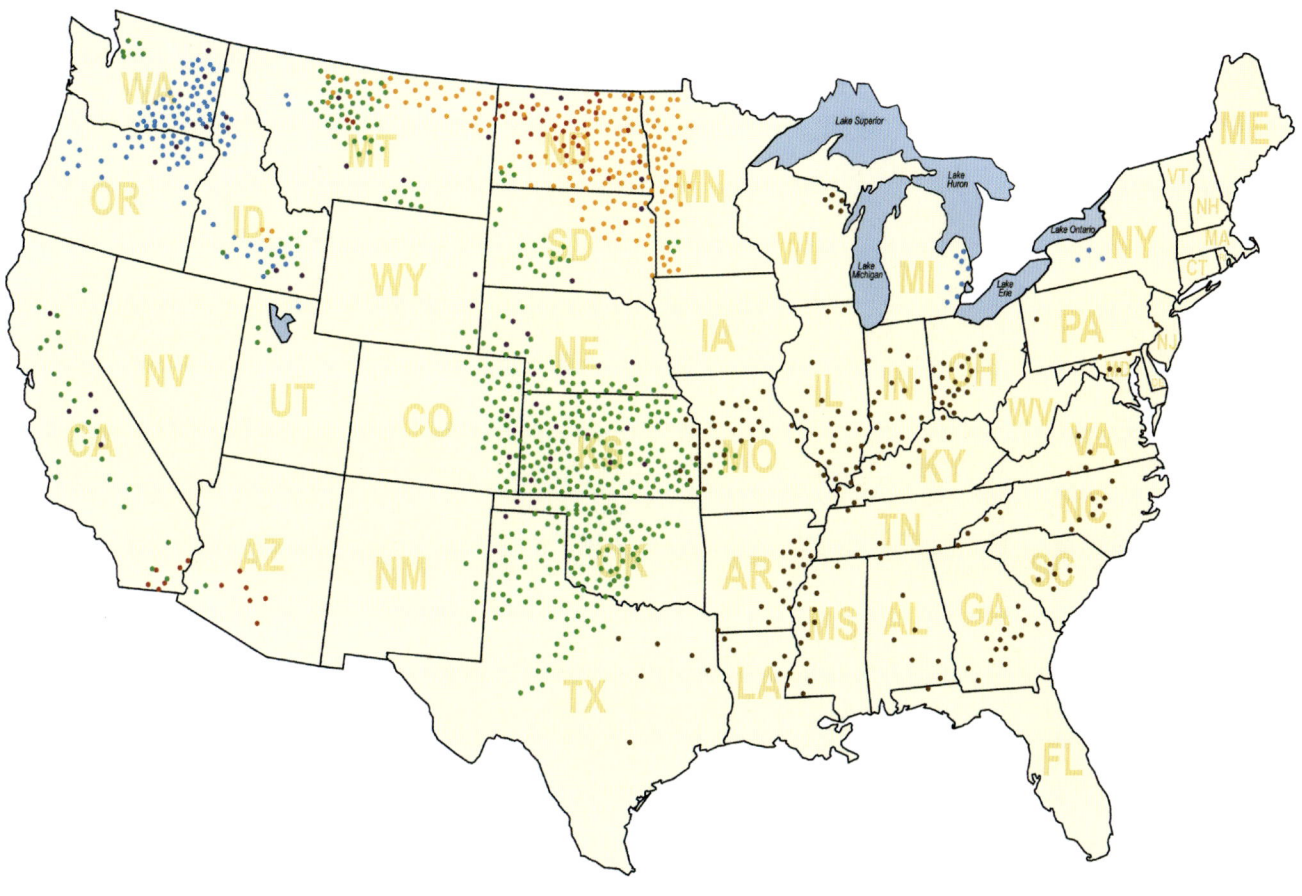

- **Hard Red Winter** (HRW) 9.5–13.5% protein content

  Bakers depend on this wheat from America's heartland. Versatile, with excellent milling and baking characteristics for pan breads, HRW is also a choice wheat for Asian noodles, hard rolls, flat breads, general purpose flour, and as an improver for blending.

- **Hard Red Spring** (HRS) 12–15% protein content

  The aristocrat of wheat when it comes to "designer" wheat foods such as hearth breads, rolls, croissants, bagels, and pizza crust, HRS is a valued improver in flour blends.

- **Soft Red Winter** (SRW) 8–11% protein content

  SRW is a high-yielding and profitable choice for producing a wide range of confectionary products such as cookies, crackers, and cakes, and for blending for baguettes and other bread products.

- **Durum** 11–15% protein content

  Hardest of all wheats, durum has a rich amber color and high gluten content. It sets the "gold standard" for premium pasta products, couscous, and some Mediterranean breads.

- **Hard White** 10–15% protein content

  As a white bran, for a wide range of first-rate products, this newest class of U.S. wheat receives enthusiastic reviews when used for Asian noodles, whole-wheat or high-extraction applications, pan breads, and flat breads.

- **Soft White** (SW) 8.5–11.5% protein content

  Exhibiting versatility and brightness for pastries to flatbreads, SW is a low-moisture wheat with excellent milling results. SW provides a whiter and brighter product for exquisite cakes, pastries, and other confectionary products.

Fig. 1. Wheat production areas and uses of six market classes in the United States (courtesy of U.S. Wheat Associates).

tremendous value of barley's value-added products. Malting barley is key to beer production, and the total business activity of the U.S. brewing industry is estimated to be nearly $200 billion. Almost 80% of the total U.S. spring malting barley production is in three states (North Dakota, Montana, and Idaho). About 98% of all Canadian barley is produced in three prairie provinces (Manitoba, Saskatchewan, and Alberta).

Oats

Oats, *Avena sativa* L., ranks fifth in world importance with 1.8 billion bushels produced in 2003 (Table 1). Almost 75% of U.S. production comes from 12 North-Central states. Oats can be "steel-cut" and then rolled, or rolled whole for processing. Steel-cut rolled oats are primarily used for quick-cooking flakes; the major portion of oats processed for food are rolled whole for use in oatmeal. Per

An Introduction to Small Grains

capita U.S. consumption of oats is ~4.6 lb, in the form of oatmeal, prepared breakfast foods, infant foods, and other food products. Oats is widely used as a feed for horses, and oats is also grown as winter forage for livestock in the southern United States.

Rye

Rye, *Secale cereale* L., is the only small grain that is cross-pollinated. It was first used in North America as a green manure crop, but now winter and spring types are grown for grain. More than 90% of the world's rye crop is grown in Europe and Russia. It ranks last in world and North American importance among the grain crops (Tables 1 and 2). Rye is mainly used in North America as temporary winter forage grazing for livestock. Rye has greater cold hardiness and usually provides better winter forage growth than other small grains.

Triticale

Triticale, *X Triticosecale* Wittmack, is a relative newcomer to North American agriculture. It is produced by crossing wheat and rye. Most triticales are hexaploid, by a cross of tetraploid wheat and diploid rye. Triticale is an attempt to combine the improved yield potential and grain quality of wheat with the improved cold hardiness, vigor, and pest resistance of rye. Currently, triticale is widely grown in northern Europe, but it is not grown extensively in North America. Improved varieties have generated interest in triticale for use as temporary winter forage grazing for livestock and as a winter cover crop. Triticale is more important in certain European countries as an animal feed.

Selected References. 1, 19, 40, 160

By David R. Porter

Growth and Development of Small Grains

The cereal plant uses water and mineral nutrients from the soil and carbon dioxide from the air to grow and produce grain. Managing wheat for optimum yields requires that certain practices, such as nitrogen fertilization and application of pesticides, be timed at specific stages of wheat growth. Correct identification of growth stages is critical for in-season management decisions about fertilizer applications, herbicide and insecticide selection and timing of application, and harvest scheduling.

Growth and development are related, but separate, plant processes. Growth often is described as an increase in size or dry matter, whereas development involves differentiation into tissues and organs. Growth rate is determined by many factors, including genetics, soil type, soil fertility, planting depth, planting date, water availability, and planting density. Development rate is determined primarily by temperature, photoperiod, and crop class. Small grains grown in North America are cool-season plants that sustain growth rates at lower temperatures than warm-season crops, such as corn or grain sorghum.

Growth

Water and mineral nutrients are absorbed from the soil by the roots of the grain plant. Water and nutrients move through the plant in the vascular system. The plant retains only a small amount of the water that is taken up. The remainder evaporates through pores called stomata in the surface of leaves and stems. This evaporation process, transpiration, occurs when stomata open in response to light and sufficient soil moisture so that the plant can absorb carbon dioxide from the air.

Chlorophyll and other pigments in the green plant trap light energy to make sugars from carbon dioxide and water. This process is called photosynthesis. The products of photosynthesis are used for the plant's energy needs and as building blocks for growth and grain production. Stresses that decrease photosynthesis cause smaller leaves, stems, and heads, and lower grain yields. Stressed plants compensate by sacrificing total growth but still go through each stage of development.

Development

The pattern of development of the wheat plant can be used to guide crop management. Small grain development is determined mainly by crop class (winter vs. spring) and temperature. Photoperiod also may be important in photoperiod sensitive genotypes. Winter type crops often have a vernalization period of varying length, whereas spring types typically do not require vernalization. Plant development of wheat is driven by temperature, which can be timed by growing degree days (GDD). Wheat grows and develops when the average daily temperature exceeds 32 °F.

The relationship between small grain development and temperature can be determined by the number of wheat GDD. GDD are calculated by the formula $GDD(°F) = (max + min / 2) - 32$, where max is the maximum daily temperature, and min is the minimum daily temperature. Temperatures <32 °F are recorded as 32 because this is the lowest threshold for growth and development. The upper developmental limit is ~95 °F, so temperature above this level should be entered as 95. Celsius GDD can be calculated by converting degrees F to degrees Celsius (°C) = $5/9$ (°F–32).

An Introduction to Small Grains

Yield

Grain yield (kernel weight per unit area) is a function of genetics and environment. Yield of a given variety is dependent on plants per unit area, tillers per plant, grain heads per tiller, kernels per head, and weight per kernel. The factor most directly associated with yield is kernel number per unit area, which is dependent upon tiller production, head, and seed development. Small grains can compensate for injury during earlier stages of plant development, but usually they cannot compensate for direct injury in the final stages (dough stages to maturity) of grain development.

Growth Stages of Small Grains

Germination and Seedling Growth. Wheat germinates between 39 and 90 °F, but optimal germination occurs between 68 and 77 °F. Germination is indicated by radicle (primary root) protrusion through the seed coat and followed by emergence of the coleoptile (first leaf), which surrounds and protects the emerging stem and primary leaves. Germination normally is completed within 4–6 d at optimum temperatures. All small grains require ~140 GDD to complete germination and ~90 GDD to emerge from each inch of planting depth. Dry soil conditions delay germination and increase GDD requirements for emergence.

Wheat produces a root and crown (nodal) system that develops sequentially according to a pattern typical of grasses (Fig. 2). The wheat plant has seminal and crown (nodal) roots. Seminal roots usually have 3–6 main roots and their branches. The first seminal root to appear is called the radicle, which is produced at germination. All other seminal roots arise from the nodes. Variety and seed size are the main factors affecting the number of seminal roots.

Crown roots are produced on main stems, primary tillers, and secondary tillers. Each main stem node develops two roots after the leaf first appears. If a tiller is produced at the node, one and sometimes two roots elongate after the first appearance of a leaf at the node.

Seedling growth occurs from coleoptile (first leaf) emergence to tiller development. Generally, the wheat plant develops three or more leaves before tillering. The rate of individual leaf growth, and the final shape and size of the leaf are affected by the environment. During vegetative growth, wheat can be distinguished from other small grain crops by its short hairy auricles, which are located at the point where the leaf blade and sheath meet (see Fig. 3).

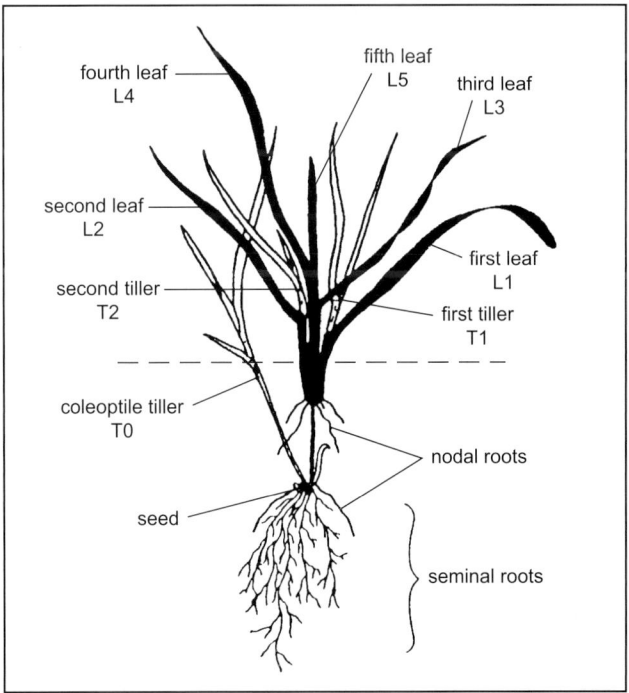

Fig. 2. Anatomical structures of a small-grain seedling.

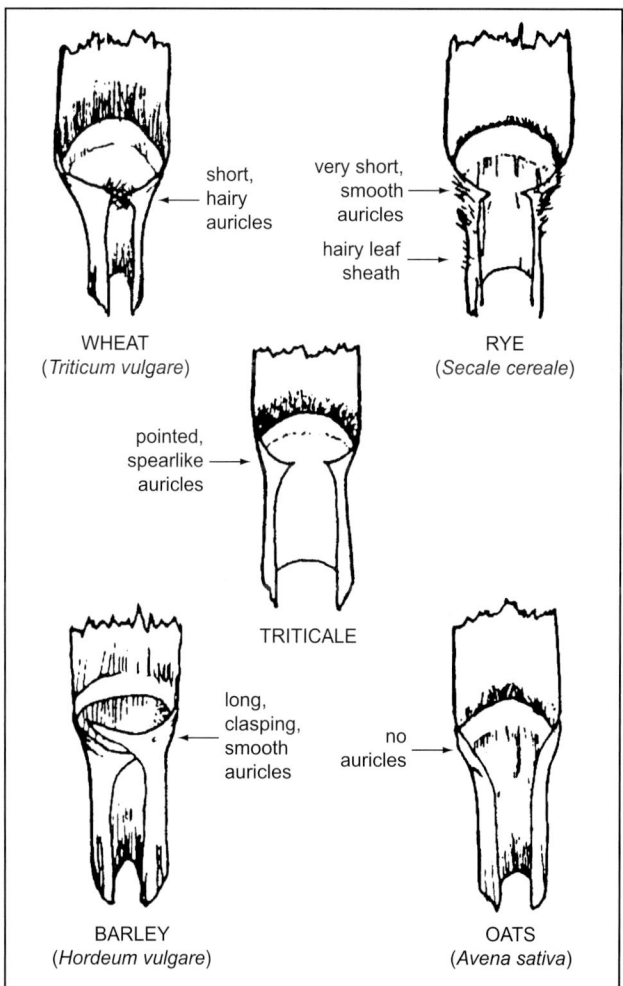

Fig. 3. Leaf characteristics of wheat and other small grains.

An Introduction to Small Grains

Different small grains develop leaves at different rates. For example, winter wheat requires 180 GDD(°F) to develop each main stem leaf (this developmental interval is termed a phyllochron), whereas spring wheat requires ~144 GDD and spring barley requires 126 GDD. GDD requirements also vary among varieties of the same crops.

Tillering. Tillering is the development of shoots from buds at the base of the main stem (Fig. 2). Every leaf on a cereal plant has a tiller bud at its base. The first primary tiller begins developing when leaf 4 emerges, the second primary tiller when leaf 5 emerges, and the third when leaf 6 emerges. Successive tillers develop one fewer leaf than the tiller preceding it, such that heading, flowering, and grain development occur at approximately the same time on all tillers of each plant. The number of leaves on the main stem provides a measure of plant development. From planting to 6-leaf stage on the main stem (3 tillers) requires ~1300 GDD.

Generally, crown roots begin to grow from a tiller after it has at least two leaves. After a tiller has three emerged leaves, the length of roots increase. The depth of rooting increases directly with root number and is also influenced by the soil profile. When wheat reaches jointing or early boot stage, new root production ceases and growth continues in the existing roots. Each tiller has roots that first appear when the tiller has 2–3 leaves.

During initial development, the tiller is dependent upon the main shoot for nutrition; but once the tiller develops ~3 or more leaves, it becomes independent of the parent plant for nutrition and forms its own roots. Varieties show relatively little variation (5–10%) in tiller development rate; planting date and season-to-season climate have a greater effect on leaf development rate than does variety.

Secondary tillers arise from primary tillers. The extent of secondary tillering is dependent upon varietal genetics and environmental factors. Tillering increases with high light intensity, reduced plant populations, and the availability of high soil nutrient (primarily nitrogen). High temperatures, dense plant populations, soil moisture stress, and pests may reduce tillering. Although each tiller has the potential to bear a productive seed head, generally, about one-half of the tillers produced do not survive to bear grain. Abortion of tillers occurs early in development, long before they are visible. However, if a plant that has been stressed during vegetative development is exposed to a favorable environment, rapid tiller growth can improve grain yield. The relative number of tillers is an indication of whether or not a plant developed under stressed conditions.

Vernalization. The onset of reproduction is controlled by vernalization, i.e., the induction of the flowering process by extended exposure of the shoot apex to low temperatures. Vernalization can occur in seeds as soon as they absorb water and swell. The vernalization requirements for winter small grains ranges from 1 day to 10 weeks. Spring type small grains essentially have no vernalization requirement; the onset of reproduction occurs after the accumulation of a required number of GDDs.

The effectiveness of vernalization declines with increasing plant age. Vernalization is affected by photoperiod; plant exposure to short days replaces the requirement for low temperatures in some varieties. Also, if wheat is exposed to high temperatures (86 °F [30 °C]) shortly after low temperatures, vernalization will not occur. After vernalization, the initiation of flowering may be hastened by longer photoperiods because wheat is a long-day plant (requiring shorter nights). Generally, early maturing varieties require fewer chilling hours of vernalization than late maturing ones.

Jointing and Grain Head Formation. The grain head differentiates before the main stem and tillers begin to elongate. In wheat and oats, formation of new spikelets ends with the formation of a terminal spikelet, usually by the appearance of the sixth leaf; barley does not form a terminal spikelet. After all spikelets have formed, seasonal nitrogen applications and other cultural practices no longer affect the potential kernel number, although supplemental nitrogen may increase the number of filled kernels, kernel weight, and final grain protein content.

Stem elongation or jointing begins when the first internode of the stem is visible. Generally, wheat stems will have 5–6 internodes, and the internodes increase in length from the base of the plant to the top. Stem height is under genetic control, but expression is affected by environmental conditions. Jointing ends at the appearance of the uppermost leaf, called the "flag" leaf, at ~2400 GDD(°F).

Reproductive development is first observed in the "boot stage" when the head begins to swell within the flag leaf sheath. The spike (grain head) is composed of rows of spikelets on the terminal end of the last stem internode (rachis). Each spikelet produces 2–5 florets, and each floret may produce a single grain. The number of spikelets depends on environmental conditions during early jointing. High temperatures increase the rate of spike development but reduce the number of spikelets per head; moisture stress reduces the number of spikelets. High light intensities and optimal nitrogen fertilization increase the number of spikelets.

The boot stage ends and heading stage begins when the spike first emerges from the flag leaf sheaf. Pollination occurs after the grain head emerges from the flag leaf. Small grains are normally self-pollinated. Rye is cross-pollinated. Pollination begins in the middle region of the head and progresses to the tip and base. Most pollen is shed before anthers emerge from florets. Barley typically sheds its pollen at about the time that heads emerge from the boot, whereas wheat and oats generally shed pollen 2–4 d after heads have emerged from the boot.

Grain development begins with pollination (joining of pollen and egg). High temperatures and drought stress during heading can reduce pollen viability and reduce the number of grains. Conversely, freezing temperatures may cause head injury and partial or complete sterility during jointing and heading. Few differences among varieties have been found for cold damage during this growth stage.

Grain Filling and Ripening. A few days after pollination, starch begins to accumulate in the kernels, and accumulation continues until kernels reach physiological maturity. Within the plant, developing spikelets compete for limited supplies of photosynthate and nitrogen. Florets at the tip of each spikelet are often unable to obtain enough nutrients to keep growing. Some spikelets at the base of wheat and barley heads also may fail to develop. The number of florets that develop is often less than the number of primordia formed before jointing.

Starch and protein are the primary storage reserves in the mature kernel. Most of the carbohydrates are accumulated in the kernels from photosynthetic output of the flag leaf. Consequently, any stress or pest injury that limits size or function of the flag leaf has pronounced effects on grain yield and quality. Environmental factors, primarily high temperature and moisture stress, affect kernel survival and the rate and duration of grain development. Starch deposition within the grain is under greater environmental influence than protein accumulation. With high temperatures and moisture stress, there is less starch concentration and final grain dry weight.

Small grains are physiologically mature in the hard dough stage, which occurs when the grain is no longer soft and pliable. The moisture content may range from 25 to 35%. The entire plant then loses chlorophyll and assumes a characteristic straw color. At this point, the crop is ready for harvest at 13–16 % grain moisture content.

Development Scales

Several small grain development scales have been developed. The two scales most frequently used are the Feekes and Zadoks scales (see Table 3, next page). The Feekes scale provides a numerical system for describing wheat growth, but is not very specific during the germination, seedling, jointing and booting stages. The Zadoks scale is based on a two digit descriptive system, which allows for more detail in quantifying wheat development. These growth scales allow for comparisons of development among varieties in varying environments. They also can aid the proper timing of management practices such as nitrogen fertilization and pesticide treatments.

By Jerry W. Johnson and Larry Robertson

Small Grain Production Practices

Wheat is widely adapted with measurable acreage in each state in the continental United States and in most Canadian provinces. Other small grains (rye, barley, oat, triticale) also are widely grown using similar production practices, but on a much smaller scale. Production practices change depending on the location and intended use of the crop. Most of the wheat is winter wheat, except in far northern regions where spring wheat is planted, and far southern areas where facultative wheat is planted. Facultative wheat does not require vernalization to convert from vegetative to reproductive growth as does spring wheat, but facultative wheat is not killed by cold temperatures, as spring wheat would be.

Most small grains are grown for grain only. Some noted exceptions are rye, which is frequently planted as a cover crop and winter forage; oat that is occasionally used for hay, and wheat, which is often used for dual purpose in the southern Great Plains and Southeast. Dual-purpose wheat is grazed during at least part of the vegetative period; and the cattle are removed before jointing, after which the wheat plant develops to maturity for grain harvest.

Cropping System

Small grains are grown without crop rotation in many areas of the Great Plains. In this region, continuous wheat fields may be planted to wheat every year or may be fallow every other two years. In a wheat–fallow system, fallowed land stores water to improve the potential yield of the sequential wheat crop. In drier areas, a common three-year rotation with two crops is wheat followed by a fallow period until sorghum is planted the following May, and then another fallow period before returning to wheat. It is difficult to rotate winter wheat with summer crops in the same year in this area because late harvest of the summer crops decreases the probability of successful fall establishment of wheat. Depletion of soil moisture and low potential for rainfall after planting wheat also are a problem.

In the remainder of the United States, small grains are frequently rotated with other crops. A common rotation in the East, Midwest, and mid-South United States is three

Table 3. Description and Comparison of Zadoks' and Feeke's Wheat Development Scales

Zadoks' Scale	Feeke's Scale	General Description	Additional Remarks
		Germination	
00		Dry seed	
01		Start of imbibition	
03		Imbibition complete	
05		Radicle emerged from caryopsis	
07		Coleoptile emerged from caryopsis	
09		Leaf just at coleoptile tip	
		Seedling Growth	
10	1	First leaf through coleoptile	Second leaf visible (<1 cm).
11		First leaf unfolded	
12		2 leaves unfolded	
13		3 leaves unfolded	
14		4 leaves unfolded	
15		5 leaves unfolded	50% of laminae unfolded
16		6 leaves unfolded	
17		7 leaves unfolded	
18		8 leaves unfolded	
19		9 or more leaves unfolded	
		Tillering	
20		Main shoot only	
21	2	Main shoot and 1 tiller	
22		Main shoot and 2 tillers	
23		Main shoot and 3 tillers	
24		Main shoot and 4 tillers	
25		Main shoot and 5 tillers	
26	3	Main shoot and 6 tillers	
27		Main shoot and 7 tillers	
28		Main shoot and 8 tillers	
29		Main shoot and 9 or more tillers	
		Stem Elongation	
30	4–5	Pseudo stem erection	
31	6	1st node detectable	Jointing stage
32	7	2nd node detectable	
33		3rd node detectable	
34		4th node detectable	Nodes above crown
35		5th node detectable	
36		6th node detectable	
37	8	Flag leaf just visible	
39	9	Flag leaf ligule/collar just visible	
		Booting	
40		—	
41		Flag leaf sheath extending	Early boot stage
43		Boots just visibly swollen	
45	10	Boots swollen	
47		Flag leaf sheath opening	
49		First awns visible	In awned forms only
		Inflorescence Emergence	
50	10.1	First spikelet of inflorescence	Just visible
52	10.2	1/4 of inflorescence emerged	
54	10.3	1/2 of inflorescence emerged	

Table 3. (Continued)

Zadoks' Scale	Feeke's Scale	General Description	Additional Remarks
56	10.4	3/4 of inflorescence emerged	
58	10.5	Emergence of inflorescence completed	
		Anthesis	
60	10.51	Beginning of anthesis	
64		Anthesis half-way	
68		Anthesis complete	
		Milk Development	
70	—		
71	10.54	Caryopsis watery ripe	
73		Early milk	
75	11.1	Medium milk	Notable increase in solids of liquid endosperm
		Dough Development	
80	—		
83		Early dough	
85	11.2	Soft dough	
87		Hard dough	Fingernail impression not held
89		inflorescence losing chlorophyll	Fingernail impression held
		Ripening	
90	—		
91	11.3	Caryopsis hard	Seed difficult to divide by thumbnail
92	11.4	Caryopsis hard	Seed no longer dented by thumbnail
93		Caryopsis loosening in daytime	
94		Overripe, straw dead and collapsing	
95		Seed dormant	
96		Viable seed giving 50% germination	
97		Seed not dormant	
98		Secondary dormancy induced	
99		Secondary dormancy lost	

crops in a two-year cycle of winter wheat followed by soybeans and then corn in the second year. Other winter cereals also can be grown with this system. In the northern Great Plains where spring cereals are grown, annual rotations with crops such as corn, soybeans, canola, and sunflower are easily established. Other common rotations include forages (grass and legumes) and rotations with peas.

Cultivar Selection

In cultivar selection, it is important for producers to consider grain yield, test weight (grain bulk density), maturity, plant height, susceptibility to lodging, disease resistance, and insect resistance. Selection for disease resistance will change in priority depending on the history of disease severity at each location. Some of the diseases to consider are leaf rust, stem rust, Septoria and Stagonospora leaf and glume blotches, tan spot, powdery mildew, and viral diseases including soilborne wheat mosaic, wheat spindle streak mosaic, wheat streak mosaic, and barley yellow dwarf. Insects of importance and those for which resistant wheat are available are discussed later in this book. Additional characteristics used in variety selection could be winter hardiness, forage production capability, aluminum (pH) tolerance, drought tolerance, and coleoptile length. End-use quality may also be an important characteristic in selecting cultivars, particularly for barley that is used for malting.

For wheat, a choice may also need to be made between a hybrid or pure line. Currently, most of the wheat acreage is planted to pure-line cultivars, but hybrids are being used in some high-yielding environments.

Seedbed Preparation

Small grains can be successfully produced in all tillage systems. Regardless of tillage system, a firm, moist seedbed is critical for rapid uniform seedling emergence. Good seed–soil contact in a moist seedbed is more important than the tillage system for good stand establishment.

An Introduction to Small Grains

Reduced or limited tillage systems where crop residue remains on the soil surface generally retain soil moisture and reduce soil temperature. This may be an advantage for winter wheat planted early where grazing is emphasized; however, lower temperatures under limited tillage conditions may delay emergence of spring-planted crops. In the High Plains, leaving previous crop residue on the soil surface has improved water conservation and increased wheat yield. In the Central Plains, leaving mulch for water conservation has not consistently increased yield.

Tillage is used to break up compacted soils and to control weeds or volunteer crops. Where *Bromus* spp. is a problem weed, moldboard plowing is promoted to bury *Bromus* spp. weed seed. However, soil conservation benefits, reduced input cost, and farm program conservation compliance requirements have increased the number of acres of mulch/reduced tillage across the small grains–producing region in recent years.

Fertilization

Nitrogen is the most common limiting nutrient in small grain production. The nitrogen requirement generally ranges from 1.5 to 2 lb (0.68–0.91 kg) of actual nitrogen per bushel of grain yield expected. Nitrogen management depends on soil type and rainfall. In deep sandy soils, multiple applications should be used to prevent nitrogen from leaching below the root system. In the drier regions of the Great Plains, it is common to apply all of the nitrogen before planting. In higher rainfall areas, the nitrogen may be split into several applications. For example, 30% of nitrogen would be applied before planting and 70% applied just before jointing; or 30% of nitrogen would be applied before planting, 40% applied just before jointing, and the remaining 30% applied during the boot stage. A soil test is needed to determine whether phosphorus or potassium fertilizers are needed. Soil test for nitrogen is a good management tool in small grain areas with low rainfall; but in higher rainfall areas, soil testing for nitrogen is not as reliable because of the mobility of nitrogen in the soil.

The wheat production areas in the United States, Canada, Australia, South Africa, and parts of South America have soils that are naturally low in pH. Nitrogen fertilization also lowers soil pH. When soil pH is low enough to reduce the production of small grains (usually <6.0 pH), the increasing level of soil acidity increases the availability of aluminum and/or manganese and causes aluminum or manganese toxicity. Acidic soils normally are amended periodically with limestone to raise soil pH. Small grain species and cultivars within species respond differently to aluminum toxicity (i.e., pH). Rye is the most tolerant small grain to aluminum toxicity followed by wheat, oats, and barley. Within wheat, cultivar tolerance to aluminum toxicity varies greatly. In areas with low soil pH, aluminum tolerance is a selection criterion in several wheat-breeding programs.

Row Spacing

Small grains are planted in narrow rows with a grain drill. Row spacing historically has been 7–16 in. (18–41 cm); however, recent research in the southern Great Plains has shown that wheat yield increases when row widths were narrowed to 3 in. (8 cm). Most producers use row spacing of 6–8 in. (15–20 cm). Wider rows may be used to allow for greater trash clearance while drilling in minimum tillage conditions or to reduce soil movement across adjacent rows when planting with a deep furrow drill in dryland conditions. A deep-furrow drill is used when rainfall is erratic, and seed placement in moist soil 2–4 in. (5–10 cm) below the surface is needed. The deep-furrow drill also leaves undulating soil surface to help trap snow for insulation.

Weed Control

Weed control is accomplished in several ways in small grain production. In limited tillage systems, herbicides are used to kill weeds present at planting; and preemergence or post-emergence herbicides may be used in the wheat crop. Tillage can be used to eliminate weeds before planting, but a herbicide may still be used during the season. Crop rotation and delayed planting also may be used to reduce weed problems in small grains. For example, delaying wheat planting and using the last tillage operation to destroy newly germinated cheat grass has been effective in reducing the competition between cheat and wheat. Herbicide-resistant cultivars of wheat also may be commercially available in the future.

Foliar Fungicides

Foliar fungicides can be used on wheat to increase grain yield and quality when foliar diseases such as leaf rust, tan spot, Septoria leaf and glume blotch, or powdery mildew are present. Yield increases as high as 39% have been obtained on susceptible varieties when application has been timely and disease pressure was high. Increases in grain and flour protein content of 0.7% have been obtained with foliar fungicide as compared with untreated grains.

Harvest

Wheat, rye, and triticale are usually harvested directly with a combine after grain moisture has reached a safe level to store the grain (<13%). Grain can be harvested when moisture is as high as 18% and dried mechanically. Barley and oat may also be directly combined, but they often are swathed into windrows and picked up later to reduce shattering losses. Harvest usually begins with winter wheat in May in the South and ends in September with spring wheat and barley in the Northern Plains. Extended rainfall after crop maturity can cause serious harvest losses due to lodging, grain sprouting in the head, lowered test weight, and interference from rapidly growing annual weeds. Test weight is very important in small grains because it is routinely used as a quick measurement of grain quality. Low test weight typically reduces the grade and price received (sometimes substantially) for the grain.

Winter Grain Production

Planting Date. The optimal planting date for winter small grains usually is September or October in the Midwest and southern Great Plains regions; the farther north in this region, the earlier the planting date. However, optimal planting date in the Gulf Coast and southeastern United States can be as late as the second half of November. Planting earlier than the optimal time often results in reduced yields caused by several factors including increased severity of infestation by Hessian fly and aphids transmitting barley yellow dwarf virus and increased infection levels of root rot and foliar diseases. Early plantings also may bolt during warm periods in the winter and therefore are more susceptible to winter kill. Conversely, late plantings do not tiller enough to maximize yield. In all cases, the objective is to have a wheat plant with 3–4 tillers emerged before winter dormancy. Such plants have developed a good crown root system and seem to survive cold stress better than smaller or larger plants.

Major exceptions to these planting dates apply in drier areas where the small grain is planted anytime in the fall when enough moisture is available to cause germination. This is especially true in the Western Plains where fall precipitation is erratic. Another exception occurs in double cropping when the small grain would be planted as soon after harvest of the summer crop as possible, but this frequently is later than the ideal planting date.

Seeding Rate and Depth. Seeding rates vary for several reasons. Generally, for an ideal planting date in a grain-only system in the Great Plains, a seeding rate of 1 bushel per acre is probably most common. In high rainfall areas, producers seeking high yields use a seeding rate of 2 bushel per acre. In low rainfall areas, rates of 1/2–3/4 bushel per acre may be used. However, seed size can vary considerably among small grain cultivars. More recently, seedling rates have been based on seeds per unit area or row length instead of bushels per acre to ensure an optimal number of plants per unit area.

Seeding depth most commonly used is 1–1 1/2 in. (2.5–3.8 cm). When surface soils are dry, seeds may be placed as much as 3 in. (7.5 cm) deep. This results in excellent stands as long as no rain occurs before emergence or the soil temperature and variety combination results in a coleoptile long enough to reach the soil surface. If a rain occurs and the soil crusts, deep planting may result in stand failure when the coleoptile does not reach the soil surface.

Spring Grain Production

Spring grains include most of the oats planted in the United States, as well as barley and wheat in the northern Great Plains and into Canada. They are planted in this area in the spring because of their inability to survive the winter.

Planting Date. Spring small grains should be planted as soon after January as planting can be done. Where the soil freezes, planting should begin after thawing and when moisture is correct for planting. In most situations, yield drops rapidly if planting occurs after the earliest planting window.

Planting Depth and Rate. Spring planting of small grains is usually done ~1 in. (2.5 cm) deep and at 2 bushel per acre. Higher seeding rates are used than for fall plantings because of the shortened season and reduced time for tillering.

Small Grains for Forage

All small grains can be used for forage. They may be used for dual purposes as mentioned earlier, for full-season grazing, for a hay crop, or simply for a cover crop for soil conservation or green manure. Straw also is an important secondary product of small grain production near urban areas. Small grains specifically for grazing by cattle often are blends of several small grains including rye, oats, wheat, and/or ryegrass.

When small grains are used for dual-purpose or full-season grazing, producers frequently are interested in maximizing forage production in the fall and winter. To accomplish this goal, planting is done 1 month or more earlier than the ideal date for grain production and seeding rates are raised to as high as 2 bushel per acre.

Planting depth also becomes much more critical for two reasons. First, as soil temperature increases, the maximum potential length of the seedling coleoptile decreases.

Therefore, the seed needs to be planted shallower to ensure that the coleoptile reaches the soil surface. If the coleoptile does not reach the soil surface, when the first leaf emerges from the tip of the coleoptile; it becomes trapped in the soil. Second, because planting is done in much hotter soils and there is concern that the soil will dry before germination occurs, producers have a tendency to plant deeper. However, planting shallow and expecting that some of the seed will not emerge until it has rained usually is best.

Selected References. 33, 56, 74, 77, 87, 177, 193, 203

By Eugene G. Krenzer, Jr.

Insect and Mite Injury to Small Grains

Successful pest management strategies for small grain insects rely on an accurate quantitative assessment of the pest population combined with a detailed knowledge of the crop damage potential. Contemporary approaches emphasize the development of IPM programs, which use combinations of control tactics holistically. The central rule for IPM is to establish an economic injury level (EIL) that accurately predicts when a particular control tactic should be applied to a pest population. The EIL, defined as the minimum pest population density that will cause economic damage, provides the basis for establishing the economic or action threshold, which is the level at which direct control tactics are warranted. Continuous variation in market values, production costs, and management tactics render EILs that can fluctuate tremendously from year to year. In this context, the progressive development of IPM requires a detailed knowledge of insect–plant–environmental interactions with particular regard to plant injury. The assessment of insect-induced injury provides the basis for developing a yield- or quality-loss relationship that governs the control tactic decision-making process.

Types of Injury

More than 120 species of insects and mites are reported to cause economic losses in small grains. Pest infestations can occur from the time of seeding, throughout all plant growth stages, until the grain is consumed as food or feed. All parts of the plant are subject to attack. Arthropods that attack cereals can be grouped into chewing, boring, and sucking pests that primarily cause crop damage by consuming plant tissue, removing plant fluids, and transmitting pathogens, as well as the predisposition of plants to biotic and abiotic stresses.

Yield losses to arthropod pests can result from either direct or indirect injury. Direct injury occurs when insects feed directly on cereal grains. Indirect injury results from feeding on roots, stems, or leaves, which are physiologically related to grain yield but do not produce seed themselves. Direct injury is typically measured by sampling techniques that estimate either absolute or relative amounts of damaged grain and, consequently, has a strong correlation with yield loss. In contrast, indirect insect injury is commonly assessed by the relative amount and character of the extant visible damage. The relationship between indirect damage and yield loss is complex; and it can be moderated by many physical, chemical, and biotic factors, making it much more difficult to assess. For direct and indirect pests, the amount of damage is directly related to the pest density and the duration of the infestation and can be differentially affected by numerous environmental interactions. The character of damage inflicted by insects largely depends upon the plant part that is attacked and the feeding strategy used. Together, these factors govern the probability and the extent of yield loss to the crop.

Plant Response to Injury

Plant injury occurs when stress-induced aberrant metabolism results in irreversible physical or chemical changes. Often, plant injury is expressed as reductions in growth, grain quality, or crop yield. Therefore, in an IPM context, understanding how insect stresses influence the expression of plant injury is critical to developing accurate EILs. To accurately assess plant injury, it is important to understand the nature of the underlying plant stress. The complex responses of cereal plants to stresses imposed by arthropod infestations involve metabolic alterations that invoke a broad range of developmental and cellular events, many of which can be moderated by changes in the timing of infestation, the duration of the infestation, concomitant environmental stresses (particularly drought), and the nutritional status of the crop. Because individual plants function as integrated units, stress responses cannot be fully evaluated except in the context of the whole plant. Most studies focus on responses of a single physiological plant process to a specific pest over a short period of time; and little attention has been given to the effects of interacting stresses, which are concomitant with arthropod infestations.

The physiological dysfunctions underlying the damage responses of cereals to insect infestation are characteristic for the feeding strategy of the insect. For example, chewing insects that consume aboveground plant tissue can significantly reduce leaf area and directly decrease photosynthetic capacity. Root- and stem-feeding insects decrease photosynthetic rates indirectly by reducing water uptake, leaf turgor, transpiration rate, stomatal conductance, and CO_2 uptake. Moreover, decreased root densities can significantly diminish nutrient uptake and mineral transport.

Sucking insects that feed in the phloem remove photoassimilates from the plant and can alter its carbon allocation processes, thereby causing source–sink imbalances. An imbalance between source and sink can result in decreased energy reserves and predisposition to winterkill, a diminished ability of the plant to adjust osmotically, and significant reductions in grain filling.

The timing of the insect infestation in relationship to the stage of plant development is important in plant injury. Wireworms, false wireworms, armyworms, and cutworms attack and destroy germinating and newly emerged seedlings and can cause serious losses during stand establishment. However, cereals have an exceptional capacity to compensate for stand losses that occur during the seedling stage. This is readily evident in winter wheat that endures grazing stress during the fall. Stand loss is fully compensated by the increased production of basal shoots that form additional tillers in the spring. Insect infestations in combination with drought stresses are especially critical during cereal reproductive development. This is especially true at the time of anthesis and grain filling. Aphids, particularly Russian wheat aphids and greenbugs, pose the greatest threat to cereals from tillering through heading stages. Insects like the Hessian fly, can have a significant impact during the heading stage of cereals by causing premature death of plants, loss of heads, and lodging.

Selected References. 37, 39, 78

By John D. Burd

Small Grain Pest Management

Principles of Small Grain Pest Management

The management of insect and mite pests of small grains can be divided into four major strategies: cultural, plant resistance, biological, and chemical. These strategies and associated tactics are integrated into a management program with the goal of reducing the economic losses caused by insect and mite pests. The level of integration is dependent on the knowledge base available to the producer or the crop consultant, economics associated with the production of the crop, the level of risk the producer is willing to assume, and the cropping system that the producer is using.

Management strategies can be classified two ways based on how they are used. Preventive management strategies are used to prevent economic losses without knowing the level of insect pest density and usually before the occurrence of the pest. These strategies are usually considered "good" agronomic practices (i.e., crop rotation) or the use of resistant plant varieties. Responsive or curative management strategies are enacted when the insect is present and implemented when the pest density reaches the economic threshold. Insecticide intervention when a pest reaches the economic threshold is a common example of responsive management strategy. Although this is a simplification of pest management, it should be helpful for the producer and consultant to consider management strategies as preventive or responsive.

Regardless of which strategy or tactic is used by the producer or consultant, correct identification of the pest and knowledge of its biology are essential. The effectiveness of responsive management strategies depends on knowledge of pest density and the relationship between density and injury and resulting damage. It is essential to understand the relationship between density, injury, and resulting damage in determining whether a management tactic or tactics will result in an economic advantage. The last step is to determine what tactic(s) should be used in preventing a pest population from reaching the economic injury level.

For many small grain insect pests, a single tactic is the keystone for the management program; however, biology of the specific pest and agronomic limitations may preclude the economic reality of a single tactic. Planting winter wheat after the first killing frost in the U.S. Midwest has been used to reduce losses to the Hessian fly, and solid-stemmed hard red spring wheat has been used to reduce losses to the wheat stem sawfly. For widespread major pests, there may be many available management tactics. Management of sporadic pests is usually limited to a few tactics.

Small grain production and pest biology are dynamic, and management strategies must be updated and adapted continually. Pests adapt to varieties and insecticides and become resistant; plant phenological stages are not constant from year to year; economic returns and fixed costs for small grain vary from year to year. Management of the insect and mite pests of small grains is constantly changing, and the producer and consultant must be dedicated to continuous education.

The following section presents current methods and techniques for producers and consultants to consider when developing a management program for small grains. Specific regional or state information is available from state cooperative extension services and should be consulted for local or site-specific information.

Selected References. 10, 149

By Michael J. Weiss

Sampling and Decision Making in Arthropod Pest Management

Sampling to assess arthropod population levels is the cornerstone of IPM. Sampling estimates pest populations on a field-by-field basis or surveys pest incidence and distribution over a large area. Control decisions in small grains should be based on accurate and current information about pest populations and about the cost of control relative to the expected loss in yield or grain quality from pest damage.

Sampling Techniques

A sampling technique is the procedure by which pest numbers are measured. Numerous techniques are available, and none is efficient at sampling all pests. Choosing a sampling technique depends on the biology and characteristics of the pest, level of desired precision, and technique cost (time and labor). Ideally, we would like to know pest numbers per unit area (an absolute estimate), but more frequently we must relate numbers to the sampling technique (a relative estimate). Sampling techniques in IPM must provide a reasonably accurate (close to the true number) and precise (repeatable) population estimate at a minimal cost. Some techniques provide accurate and precise estimates but are too time-consuming and expensive to be practical.

Most arthropod pests in small grains can be sampled by one of the following techniques: soil or aerial traps, sweep netting, vacuum netting, direct observation, and dissecting plant parts. Each technique has its advantages and disadvantages, and selection of a technique depends on the target pest's biology and the required time and cost.

Research currently is developing remote-sensing techniques to sample pest infestations. This can involve taking infrared photos of fields from aircraft or earth-imaging satellites. Changes in plant appearance can indicate plant stress or pest infestations. Remote-sensing targets ground-sampling efforts to verify and take remedial action if an infestation is present. Remote-sensing techniques may become more important in future pest management.

Sampling Program

Sampling programs determine how samples are taken in space and time to provide pest population estimates in a timely and cost-effective manner. Components include the sample technique, the sample unit, number of sample units per field, timing during the season and during the day, and the pattern of sampling. Typically, the technique is determined by previous research.

The sample unit is the size or number of units of a sample method, such as number of plants, sweeps, or length of row per sample. The stage or time during the season for sampling is when the pest is present during a susceptible stage of plant growth. Resources usually limit the number of samples per field. Generally, scouts cannot afford to spend more than an hour in each field. Enough samples should be taken to provide a population estimate with an acceptable level of precision, for example, ±20%. The pattern of sampling also depends on characteristics of the pest; some type of zigzag, S-shaped, or Z-shaped pattern is used. An effort should be made to cover the entire field, and because some pests move into a field from the edge, margins also should be sampled.

Sequential sampling methods minimize the number of samples needed for making a control decision. Samples are taken and counts tallied until the infestation can be classified with a predetermined precision level as to whether an infestation is above or below a given economic threshold (see next section). Typically, relatively few samples are needed to classify a population if it is much higher or lower than the threshold. If the population is near the threshold, classification may not be possible. In this case, sampling stops after a predetermined number of samples are taken, and the field is sampled again within several days. When done properly, sequential sampling methods can save considerable time and resources.

Sample programs typically are developed for key pests. However, numerous minor pests occasionally reach infestation levels that require control; scouts should be aware of these pests, their appearance, injury, and sampling procedures to avoid overlooking potentially damaging infestations by minor pests. Furthermore, pest managers can combine sampling efforts for weeds and diseases while sampling for insects and reduce sampling costs for all pests.

Evaluating Control Decisions

Sampling provides information on pest densities, but knowledge of current pest densities is not enough information to decide whether control action is justified. Concepts used in this decision process are the economic injury level (EIL) and the economic threshold (ET). The EIL is the pest density that will cause yield or quality loss equal to the cost of control. EIL is based on several factors, including market value of the crops, control costs, and yield loss caused per insect. In an equation form, EIL is calculated by:

$$\text{EIL} = [\text{Control costs (\$/acre)}] \div [\text{Crop market value} \times \text{yield loss per pest} \times \% \text{ control}]$$

Control costs are expressed in value (dollars) per acre (or hectare) and include the costs of the pesticide and its application. Crop market value is expressed as dollars per unit of yield such as lb, kg, or bushel. This value can be obtained from current market prices or estimated from a crop futures market. Proportion of control (%) is the percentage of reduction in a population from a particular control measure. This is assumed to be 100%; however, some pesticides may only be partially effective, in which case an estimate of percentage of control (as a proportion) should be included. Yield loss per insect is derived from research on pest damage and yield loss relationships.

The EIL changes if any of the component factors change. For example, if control costs increase, it takes more pests (actually yield loss) to justify control action, and the EIL increases. The price of several pesticides can vary by 100% or more; a low-cost pesticide would have a lower EIL than a high-cost choice. Market value is another factor that causes EIL to change frequently. As the commodity value increases, EIL declines. Conversely, as commodity prices decline, more pests and their damage can be tolerated before the loss equals the cost of control; therefore, EIL increases. In general, EILs for small grain pests are high because of relatively low commodity values of cereal grains and the general ability of small grain crops to tolerate and compensate for pest injury without much effect on yield.

The main difficulty in establishing EILs for small grain pests is in determining the relationship between pest infestation and yield loss. Where EILs based on research data are lacking, entomologists often will develop a nominal EIL based on their previous field experience with the pest and/or limited research that is available. Nominal EILs are useful decision tools because entomologists often are conservative in their assessments and thereby avoid serious pest-associated losses. Unlike scheduled or calendar sprays, nominal EILs may prevent unneeded pesticide applications when pest populations are low.

The EIL provides the break-even point at which the yield (economic) losses from a pest infestation equal the costs of control. The ET is used to make control decisions in the field. ET is the pest density at which action should be taken to prevent a pest population from increasing to the EIL. For this reason, ET is sometimes called the action threshold. The ET is a prediction: Unless management is taken at the ET, pests will ultimately cause yield losses equal to or greater than management costs.

Because the ET is predictive, determining it is a challenge. The ET often is calculated as a percentage of the EIL. For example, if the ET is 80% of the EIL, then an EIL of 10 insects per plant generates an ET of 8 insects per plant.

The EIL and ET typically are defined as some number of pests per area or plant, but occasionally ETs will be given as a number of injured plants or a percentage of plant injury. Such ETs are rarely correct because we cannot predict future injury on the basis of past injury. We can, however, predict future injury and yield loss from current measures of pest density.

The EIL and ET provide an economic basis for making pest control decisions and a rationale basis for responsive or curative pest control decisions. The EIL is useful in guiding decisions on preventive pest control when a decision to use a pest control technology must be made before the actual pest population can be known. This includes paying more for a pest-resistant variety of plant or applying a systemic insecticide at planting to prevent seedling pest infestations. The benefit of preventive control must be based on the potential losses caused by the pest and historical risk of infestation. The EIL can be used as a basis for determining whether a pest infestation is causing economic loss and provide an assessment of how much loss will occur. This assessment can be incorporated into crop management decisions for the infested crop and form the basis for future, preventive control decisions for a pest.

Selected References. 12, 80, 150, 151

By G. David Buntin and Leon Higley

Pest Management Tactics for Small Grain Arthropod Pests

Cultural Control

Cultural control tactics are changes in the agronomic production practices (altering planting or harvesting dates, tillage, variety selection) associated with the crop that will influence the pest or beneficial arthropod and result in reduced pest injury. Cultural control is usually a preventive tactic. Successful use of cultural control tactics comes from an excellent understanding of the arthropod biology and agronomy. In other words, exploiting the "weak link" in the biology of the arthropod pest is the key to the success of a cultural control tactic. These tactics usually work well when insect injury and host plant phenology are highly synchronized, or with arthropod pests that are specialists and not generalists.

Some common examples of cultural control include delayed planting of winter wheat until after a killing frost (effective against several arthropod pests including aphids, Hessian fly, and wheat curl mite); tillage to control winter annual and early spring germinating weeds and reduce grasshopper egg survival; and early swathing to reduce lodging caused by the wheat stem sawfly. Crop rotation is also a common cultural control tactic, but it is often more effective against arthropod pests that have a narrow host range and limited ability to migrate. For example, crop rotation is not effective for grasshoppers or Russian wheat aphid, but it has been successfully used for wheat stem sawfly and wheat midge.

Before the advent of synthetic insecticides, producers often used cultural control methods to mitigate arthropod pest losses. Trap cropping has been rediscovered as a preventative tactic. One example is a crop–fallow rotation system. When grasshopper eggs are deposited in the previous year's crop that will lie fallow in the current year, trap strips can be beneficial. Small grains can be seeded in strips. When plants have emerged, the remainder of the field can be tilled forcing grasshoppers into the strips. At that time, an insecticide can be applied. The advantage is only that a small amount of area has to be sprayed with an insecticide.

Another example is the early planting of a solid-stem spring wheat around the perimeter of a field to attract wheat stem sawfly females for oviposition. Solid-stem varieties are resistant to wheat stem sawfly but usually are lower yielding than hollow-stemmed varieties. Sawfly eggs will hatch, but larval survival is significantly reduced in the solid-stem wheat. A susceptible hollow-stem variety in the field interior is protected.

In conclusion, cultural control tactics for many small grain arthropod pests can be successful in mitigating losses. However, to be successful, the practitioner needs a thorough understanding of pest biology and production of small grains.

By Richard Butts

Plant Resistance

Small grain cultivars with inherited insect-resistant properties are a proven pest management tactic. Commonly after pest attack, they have a better yield than susceptible cultivars without resistant properties. Resistant cultivars are created by identifying resistance traits in plants and transferring resistance genes to existing cultivars by sexual crosses or by genetic engineering. Progeny are then assessed for desirable levels of resistance and agronomic qualities.

Insect-resistant small grain cultivars have major economic advantages. For example, wheat cultivars developed during the 1960s with resistance to the Hessian fly returned about $600 for each research dollar invested, compared with only a $5 return on the investment for the development of insecticides during the same period. Since then, 27 rye and wheat genes for Hessian fly resistance have been identified and bred into more than 60 wheat cultivars grown in the north-central and southeastern United States. Fly-resistant cultivars have decreased fly damage, increased wheat yields, and reduced insecticide applications. In all, fly-resistant cultivars have lowered production costs by more than $50 million each year.

In response to plant-resistance genes, 16 Hessian fly biotypes have developed since the first resistance genes were released. Biotypes are strains of insects that have developed virulence to insect-resistant plant genes. Hessian fly biotypes are differentiated by how various combinations of virulence genes interact with plant genes for fly resistance. High temperatures diminish the expression of plant resistance genes to some fly biotypes. Hessian fly biotype L is now predominant in the eastern, north-central, and mid-southern areas of United States wheat production, and it is virulent to all resistance genes in commercial wheats grown in these areas. Until cultivars with biotype L resistance are grown on a large scale in these parts of the U.S. wheat production, yield losses due to fly feeding can occur, especially when fly populations reach economically damaging levels.

In the wheat-producing states of the central and western plains, greenbug and Russian wheat aphid are serious pests of barley and wheat production. Certain genes identified in rye, triticale, and wheat have resistance to various biotypes of the greenbug. Wheat cultivars with different combinations of these genes are grown in Kansas, Nebraska, Oklahoma, and Texas. Germplasm with resistance to several greenbug biotypes is presently being bred into different cultivars adapted for growth in these states. When grown on a large scale, this resistance should lower production costs for insecticide application and avoid yield losses worth more than $20 million each year.

Genes from barley, triticale, and wheat originating in Afghanistan, Bulgaria, Iran, Russia, or Spain have been used to create wheat and barley cultivars resistant to Russian wheat aphid. 'Halt', the first wheat cultivar with Russian wheat aphid resistance, was produced in Colorado during 1996 on ~2.8 million acres (1.13 million ha). Insecticide applications were reduced by ~50%, saving wheat producers about $14 million. 'Stanton' is a Russian wheat aphid-resistant wheat cultivar developed by the Kansas Agricultural Experiment Station that was released in 2001. A biotype virulent to Halt and Stanton occurred in North America in 2003. Other wheat and barley cultivars are being bred from germplasm developed in Idaho, Montana and Oklahoma. (Fig. 4).

Since 1946, Canadian and U.S. scientists have developed several solid-stemmed spring wheat cultivars with resistance to the wheat stem sawfly. During prolonged periods of cloudy weather, the full expression of the solid-stem character in some resistant cultivars is diminished. This occurrence has slowed the acceptance of sawfly-resistant cultivars by producers in the past. Many of these first cultivars also lacked adequate yield and baking quality. Recently, spring cultivars have been released with sawfly resistance, medium-to-high yields, good baking quality, and disease resistance.

Genes with resistance to the cereal leaf beetle, identified in Russian wheats in the 1970s, were used to develop the cultivar 'Downy'. These resistance genes control the production of thick layers of long wheat leaf hairs that limit

Small Grain Pest Management

Fig. 4. Wheat plants susceptible (left) and resistant (right) to Russian wheat aphid.

beetle egg laying and cause beetle larvae to die after injury from the hairs. After resistant cultivars were released, several parasitic insects also played a role in reducing cereal leaf beetle populations.

Cultivars with resistance to the wheat curl mite have been developed from crosses between wheat and wild grasses and from chromosome rearrangements of wheat and rye. Unlike cereal leaf beetle resistance, greater numbers of leaf hairs are not effective against wheat curl mites. Mite resistance in commercial wheat cultivars prevents ~75% of the losses that occur normally when mites infect susceptible plants with the wheat streak mosaic virus. The value of these regained losses is worth more than $150 million annually to North American wheat producers.

Selected References. 70, 168, 175, 199

By C. Michael Smith

Biological Control

Predators, parasitoids, and pathogens are used to manage small grain insect pests, primarily by importing exotic natural enemies to suppress exotic pests and augmenting or conserving indigenous natural enemies. These natural enemies include a wide variety of generalist predators, such as lady beetles, lacewings, syrphid flies, predatory bugs, and spiders, host-specific parasitoid wasps and flies, and disease-causing fungi, bacteria, viruses, and microsporidians.

In small grains, natural enemies consistently prevent many pest populations from causing economic injury. However, a variety of complex environmental and agricultural factors can limit the effectiveness of natural enemies. Frequent use of insecticides can reduce the numbers of natural enemies in the long term through direct exposure or reduction of prey density. Under these circumstances, pest populations can rapidly increase and cause economic damage before natural enemies can reduce populations.

The importation of predators and parasitoids from Eurasia and the Middle East against the Russian wheat aphid is one of the largest biological control projects ever conducted in the United States (1986–1993). This project involved the USDA, land grant universities, state agencies, and several international organizations. Because initial establishment attempts for predators were unsuccessful, the project shifted exclusively to importing parasitoids. Several species of parasitoids are established in the western United States, and some are causing relatively high levels of mortality. The ultimate goal of this project is permanent long-term suppression of Russian wheat aphid populations.

Greenbug, bird cherry-oat aphid, and several other aphid species often are maintained below economic injury levels by lady beetles and parasitoid wasps. Of the many aphid parasitoids found in small grains, *Lysiphlebus testaceipes* (Cresson) is the most recognizable. The effect of these natural enemies on greenbug populations is considered predictable enough in Texas and Oklahoma that natural enemy thresholds are used. (A natural enemy threshold is a density of lady beetles or percentage of parasitism at which no further management actions are needed.) Producers are advised not to treat wheat fields with insecticides if there are 1–2 lady beetles (adults and larvae) per foot row of wheat, or if >20% of the greenbugs are parasitized.

Efficient approaches for estimating parasitism have been developed and allow producers to estimate minimum levels of percentage of parasitism from the frequency of wheat tillers with one or more mummified aphids.

Aphids are often an abundant food source for generalist natural enemies in small grains. Numbers of generalist predator numbers increase rapidly on aphid populations, and in turn, feed on other potential pests, often preventing outbreaks. Host-specific natural enemies also attack many nonaphid pests in small grains. Several parasitoids including *Homoporus destructor* (Say) and *Eupelmus allynii*

(French) can have a substantial impact on Hessian fly populations. The decline in cereal leaf beetle abundance can be partially attributed to several imported parasitoids including *Tetrastichus julis* (Walker), *Diaparsis temporalis* (Horstmann), and *Lemophagus curtus* (Townes). The wheat stem sawfly is occasionally controlled by its parasitoid, *Bracon cephi* (Gahan).

Disease-causing pathogens also can have an important effect on pest populations. Fungal pathogens, including *Beauveria bassiana* (Balsamo) and members of the Entomophthorales, and several viruses can suppress small grain pests. The microsporidian *Nosema locustae* (Carning) is an important disease agent of grasshoppers. Microbial insecticides, such as *Bacillus thuringiensis* (Berliner), and some viral formulations are attractive control options for suppressing pests while conserving natural enemies. However, their delayed action, lack of persistence in the environment, and high cost currently inhibit profitable use in small grains. Future advances in microbial efficacy and persistence will provide more truly integrative control tools.

Biological control may be best used in small grains by predictably incorporating natural enemy effects on pest populations into holistic pest management programs. For example, the effect of cereal leaf beetle parasitoids can be enhanced through tillage, which increases soil temperatures, altering wasp development, emergence, and synchrony with beetle populations. Currently, research is underway to validate natural enemy thresholds for greenbug management in wheat. Use of natural enemy thresholds in pest management programs to avoid unnecessary insecticide applications is an effective way to conserve natural enemies and improve biological control. If natural enemy densities are insufficient to reduce pest outbreaks, care should be taken to conserve their numbers. A small, unsprayed reservoir within a field may allow rapid recolonization of natural enemies into previously sprayed areas and prevent pest resurgence.

Using insecticides that are least disruptive to natural enemies is also an effective way to conserve natural enemies. The use of plants that are tolerant to insect feeding also will conserve natural enemies and potentially enhance their effect on several pest species. These approaches work most successfully when used within agricultural systems designed to conserve natural enemies. By using these holistic interactions fully, producers can effectively incorporate natural enemy impact into insect management decisions in small grains.

Selected References. 18, 46, 59, 62, 65, 67, 97, 99, 100, 122, 123, 143, 172, 201

By Kristopher L. Giles, Norman C. Elliott, and Tom A. Royer

Chemical Control

Insecticides are a vital component of IPM programs for small grains throughout the United States; however, most wheat acreage only receives insecticide applications during outbreaks of various pests. The severity of infestation varies from year to year and from location to location meaning that need for treatment is sporadic in most areas. Although only a small percentage of the total wheat acreage is normally treated, during outbreaks, insecticide use can be alarmingly high because of the huge amount of acreage involved.

Unlike some crops, very little insecticide is used on small grains at planting; one notable exception has been the small percentage of the acreage where seed treatments are used to prevent damage by seed-attacking pests, such as wireworms and false wireworms. Recently, new systemic neonicotinoid seed treatments have expanded the spectrum of activity to include many aboveground seedling pests, such as aphids (bird cherry-oat aphid, greenbug, Russian wheat aphid) and Hessian fly. However, in many cases, insect problems in small grains are not common or predictable enough to warrant the cost of these new preventative treatments.

In the past, the chlorinated hydrocarbons were commonly recommended for controlling outbreaks of cutworms. For several years after these products were banned, emergency exemption often had to be declared to save crops when outbreaks occurred. Recently, some of the pyrethroid compounds have received labeling for use in small grains and, thus, provide a much-needed option where outbreaks occur.

Probably the biggest change in insecticide use in small grain production occurred with the introduction of the Russian wheat aphid. This insect is capable of causing severe damage to wheat and barley and resulted in several million acres being treated with insecticides during the late 1980s and early 1990s. Much of this acreage was treated with organophosphate insecticides. Recently populations of this pest have declined, except for local outbreaks.

The low profit margins for small grains and sporadic nature of small grain insect pests often make small grains a low priority for chemical companies' development dollars and university research funds. However, when outbreaks occur, losses or treatment costs soon reach into the millions of dollars because of the large numbers of acres that can be involved in an outbreak. In most years, probably 95% of wheat grown in North America is produced without using any insecticides. Generally, insecticides are only used when pest infestations exceed economic thresholds in the field, and there is very little use of scheduled spraying or prophylactic applications. Although opportunities for improvement exist, small grain production should be held up as an example of how judicious use of insecticides in an IPM system is working very well.

By Phillip E. Sloderbeck and H. Leroy Brooks

Diseases and Arthropod Pest Management

Diseases cause considerable damage to small grains each year. Diseases caused by fungi are most important, followed by viruses and bacterial pathogens. Diseases caused by rust (*Puccinia* spp.) fungi that attack leaves and stems are the most important diseases of small grains in North America. One or more may be important each year depending on the region and cultivar susceptibility. Other important fungal diseases of the foliage are powdery mildew [*Blumeria* (syn. *Erysiphe*) *graminis* f. sp. *tritici*], Septoria and Stagonospora blotches (*Septoria* spp. and *Stagonospora nodorum*), and tan spot on wheat (*Pyrenophora trichostoma*); spot blotch and scald on barley (*Bipolaris sorokiniana* and *Rhyncosporium secalis*, respectively), and crown rust of oats (*Puccinia coronata*). Several smuts and bunts affect small grains, and Fusarium head blight or scab (*Fusarium graminearum*) is important on wheat and barley. Important root diseases of wheat are take-all (*Gaeumannomyces graminis* var. *tritici*), strawbreaker foot rot (*Pseudocercosporella herpotrichoides*), and seedling blights caused by a variety of fungi.

Arthropod Vectors of Small Grain Viruses

Insects interact directly and indirectly with other diseases, but these interactions are primarily important in relation to many diseases caused by viruses (Table 4). Three main groups of arthropods are vectors of small grain viruses.

Aphids transmit barley yellow dwarf (BYD) which is the most economically important virus disease of cereal crops worldwide. The disease affects all of the major cereal crops except rye. Yield loss estimates of 5-10% are common, but severe epidemics can cause greater losses. BYD is caused by a group of related viruses, barley yellow dwarf virus (BYDV) and cereal yellow dwarf virus (CYDV). The viruses are transmitted to plant hosts by aphids and are not mechanically or seed transmitted. A number of viral isolates have been designated with BYDV-PAV and CYDV-RPV being two prevalent isolates in North America. The typical symptoms are a reddening or yellowing of leaf tissue, which is often more pronounced at the leaf margins (Fig. 5). However, color is variable depending on plant genotype and virus isolate. Plant stunting is a common symptom usually caused by shortened internodes (Fig. 5). Severe stunting may result in delayed or failed head emergence. In the winter cereals, symptoms are often not expressed until the spring although infections may have occurred in autumn. As the plant matures the probability of inoculation decreases and the impact of the infection on the plant growth and yield is markedly decreased.

Numerous aphids have been reported to vector BYDV/CYDV, but few species have been shown to be of importance in disease epidemiology. The bird cherry-oat aphid, *Rhopalosiphum padi* (L.), corn leaf aphid, *R. maidis* Fitch, and English grain aphid, *Sitobion avenae* (F.), are important vectors in most regions. Greenbug, *Schizaphis graminum* (Rondani), rice root aphid, *Rhopalosiphum rufiabdominalis* (Sasaki), and *Metopolophium dirhodum* (Walker) are also important vectors in some areas. There may be differences in BYDV/CYDV transmission efficiency among aphid species, aphid stage, and clones or biotypes within an aphid species. Normally 24–48 hr is needed for aphids to

Table 4. Arthropod vectors of viruses occurring naturally on small grains and viruses to which small grains are susceptible in North America.

Vectors	Viruses occurring naturally	Susceptible
Aphids	Barley yellow dwarf Cereal yellow dwarf	Maize dwarf mosaic Southern celery mosaic
Leafhoppers	American wheat striate mosaic Brome mosaic Oat blue dwarf	Cereal chlorotic mottle Rice dwarf Rice stripe
Planthoppers	Rice hoja blanca	Maize rough dwarf
Cereal leaf beetle	Brome mosaic	Maize chlorotic mottle
Mites	Wheat streak mosaic High plains Barley yellow streak mosaic Agropyron mosaic	Ryegrass mosaic

Fig. 5. Reddening of leaves (top) and stunting (bottom) of wheat plants infected with BYDV/CYDV (David Buntin).

acquire the virus and another 12-14 hr is required for the infected aphid to be able to transmit the virus while feeding on the plant. Consult references listed for more details about BYD epidemiology.

The most effective control measure for BYD in winter wheat has been to adjust planting schedules to avoid having the peak aphid flights occurring when the crop is just emerging and the plants are the most vulnerable to infection. Aphid control using insecticides including systemic neonicotinoid seed treatments or foliar applications of pyrethroid insecticides also is useful in suppression BYD in some areas. Varieties with resistance to the BYD pathogen also may become available in the future.

The next most important pathogen is wheat streak mosaic virus (WSMV), which causes stunting and narrow, yellow-mottled streaks parallel to leaf veins. Entire tillers may be killed or fail to produce seed as a result of WSMV. The only known vector of the virus is the wheat curl mite, *Aceria tosichella* Keifer. The virus is found in the midgut and hindgut of the mite, and it can transmit the virus for 7–9 d after acquisition. The virus is not transmitted through eggs. Severity of the disease is directly linked to the survival and population of the mite vector on wheat. The virus requires a "green bridge" of volunteer wheat that grows in late summer and allows the mite to survive in high numbers until the next wheat crop. The mite also infests barley, rye, oats, corn, and several grasses. However, wheat is the only important host to have outbreaks of the disease. The mites are windborne and can therefore disseminate the virus up to several miles.

Wheat streak mosaic is the most serious wheat disease in the central plains of the United States. It is found throughout the northern and central United States each year and has caused losses of $35 million in a single year. The exact distribution is unknown, but the disease is found from the Pacific Northwest to Alabama and in the northeastern United States. It was recently reported in central Mexico. Wheat spot mosaic, caused by an unidentified virus-like agent that infects wheat, barley, and rye, has been found in Alberta and Ohio. Recently, high plains virus also has been identified from wheat and corn. Both viruses are transmitted by the wheat curl mite and are often associated with WSMV. Other viruses vectored by mites are Agropyron mosaic virus on wheat and barley, and yellow mosaic virus on barley.

Leafhoppers, the third group of arthropod vectors, are vectors of many viruses that cause mostly minor damage on wheat and other small grains. Several occur in North America; many related viruses and leafhopper vectors are reported in other parts of the world. The viruses are acquired during a minimum feeding period of ~15 min. The viruses are found in the midgut and hindgut and can be transmitted during insect feeding for ~7–30 d. Some of the viruses may be propagated within the insect. Oat blue dwarf virus, found from Kansas to Manitoba, is vectored by the aster leafhopper, *Macrosteles quadrilineatus* Forbes. Barley, corn, and flax are other known hosts; but mainly oats are affected. Plants are stunted with stiffened shoots, produce excessive secondary tillers, and have a blue cast to them. Losses are usually very low. The virus multiplies in the insect and is not mechanically transmitted. Another virus vectored by leafhoppers of occasional economic importance is American wheat striate mosaic virus found in the north-central United States and Canada.

Fungi and Bacteria

Insects disseminate several pathogens in a passive manner. One example is the ergot fungus, *Claviceps purpurea*. The spores infect unfertilized flowers of small grains; particularly rye; but wheat, triticale, barley, and many grasses are susceptible. The fungus produces a dense sugary exudate called "honeydew" that contains spores. Honeydew is attractive to insects, which then carry spores to flowers of neighboring plants. These spores can be disseminated by other means including splashing rain. The fungus produces black masses of tissue called sclerotia (ergots) that serve as survival structures and the source of another kind of infective spore the next year. Ergot rarely causes significant yield loss, but the sclerotia contain compounds highly toxic to animals and humans. During overwintering in the soil, the sclerotia may be consumed by a variety of insects. In particular, the ergot beetle, *Acylomus ergoti* Casey, is attracted to the sclerotia. In the southern United States, soil insects and fungi cause sufficient deterioration of sclerotia so that they do not survive the winter.

As with all cereal grains, storage insects can promote seed molding by a variety of fungi. These fungi may be parasitic or colonize the surface tissues of the grain as it matures, or they can be produced on plant debris in the field. They produce airborne spores that adhere to the grain and can survive a year or more. Many, such as *Penicillium*, *Aspergillus*, and *Fusarium* spp., have the ability to grow with as little as 14% seed moisture content. Grain-feeding insects attracted to pockets of moist grain expose the internal seed tissues where the molds grow. In addition to direct grain deterioration, many of these fungi produce potent toxins.

Wounds created by insect feeding may assist entrance of fungi and bacteria into small grains. Wireworm feeding was associated with an increase in Cephalosporium stripe of wheat. The disease, caused by the fungus *Cephalosporium gramineum*, invades roots through wounds. Bacteria require wounds or natural openings to invade plants. Infection by leaf pathogens, such as *Xanthomonas campestris* pv. *translucens*, which causes black chaff disease on wheat, rye, and barley, and *Pseudomonas syringae* pv. *coronafaciens*, which causes halo blight of oats and rye, may be enhanced by wounds caused by leaf-feeding insects.

A variety of management tactics can be used in specific situations to control arthropod vectors to prevent or minimize pathogen infection. Control of volunteer plants and grassy weeds, important to management of arthropods that cause direct damage to grains, is important to control insect and mite vectors of small grain viruses. Typically, chemical control is not an option for the virus diseases described here because it does not reduce insect populations sufficiently to reduce virus transmission. One important exception is the use in the eastern United States of neonicotinoid insecticides as seed treatments or foliar applications of lambda cyhalothrin to control aphids and thereby limits infection and secondary spread of BYDV/CYDV in wheat. Genetic resistance is used with some degree of success for the major viral pathogens such as BYDV/CYDV and WSMV. Wheat cultivars resistant to feeding of the wheat curl mite have less damage from WSMV. It is imperative that most volunteer wheat be absent at least 2 wk before planting up to 2 miles from fields to eliminate the mites. Date of planting, tillage practices, and seeding practices that promote rapid stand establishment all affect insect populations and their interaction with small grain pathogens.

Selected References. 17, 36, 43, 64, 115, 117, 118, 131, 204

By Barry M. Cunfer

Weeds and Arthropod Pest Management

Weeds can be a serious constraint to small grain production. They compete for the same resources of light, moisture, nutrients, and space required for crop growth. Weeds can adversely affect small grains by reducing grain yield and quality, contaminating grain with weed seed, and interfering with harvest operations. Weeds also may harbor insect pests or serve as hosts for pathogens and viral diseases vectored by arthropods.

Weed species infesting wheat and other cereal grains vary considerable in frequency and severity from region to region. Although >20 weed species can occur in the same field, typically only a few dominant species are a problem in a given field each year. Annual grasses with similar requirements and life cycles as wheat tend to be especially troublesome because they are difficult to control with selective herbicides. Some common grassy weeds include annual or Italian ryegrass (*Lolium multiflorum*), jointed goatgrass (*Triticum cylindricum*), cheat (*Bromus secalinus*), downy brome (*Bromus tectorum*), foxtail grasses (*Setaria* spp.), wild oats (*Avena fatua*), and many others. Common broadleaf weeds include wild radish (*Raphanus* spp.), various mustards (*Brassica* spp.), pigweeds (*Amaranthus* spp.), lambsquarters (*Chenopodium album*), wild buckwheat (*Polygonum convolvulus*), kochia (*Kochia scoparia*), thistles (various species), and many others (Fig. 6). Sometimes

Small Grain Pest Management

Fig. 6. Wild radish weeds in wheat harbor lygus bugs and flower thrips (G. D. Buntin).

volunteer crops such as common rye (*Secale cereale*), canola (*Brassica napus*, *B. rapa*), and sunflower (*Helianthus* spp.) also may be important weeds in cereal grains.

Weed–insect interactions in small grains are poorly understood. Small grains are grown in monocultures often in very large continuous acreage. Farmers attempt to remove weeds by encouraging rapid ground cover and by using broadleaf and selective grass herbicides. Herbicide-resistant wheat is not yet available, but this kind of weed control technology most likely will be deployed in North America in the near future. Extensive monocultures may encourage insect infestations by increasing the availability of small grains to pests. Monocultures also tend to reduce nectar sources and alternative host plants that are sometimes needed to attract and retain natural enemies that may help to regulate pest populations. Increased diversity of weeds or companion plants may enhance natural enemy populations thereby reducing pest infestations. However, the benefit of managing insect populations by retaining weeds to increase plant diversity has yet to be demonstrated for cereal grains in North America. Furthermore, any potential benefit in controlling arthropod pests must be weighed against the detrimental effects of weeds on crop productivity.

One approach to enhance plant diversity is to incorporate strips of perennial pasture grasses, wildflowers, and presumably weeds. Studies in Europe have shown that such strips increase numbers and diversity of ground and rove beetles in adjacent wheat fields. These strips also provide nectar sources for hover flies, parasitoids, and other natural enemies of small grain pests.

Conversely, weeds may harbor or serve as alternative hosts for small grain pests. In the southeastern United States, little barley (*Hordeum pusillum*) is the only non-crop host of the Hessian fly. This weed can harbor the Hessian fly thereby maintaining populations in the absence of wheat production. Weeds also may serve as alternative hosts for small grain disease pathogens that are transmitted by arthropod pests. Various grassy weeds are suspected of serving as sources of viruliferous aphids that transmit barley yellow dwarf and cereal yellow dwarf viruses to winter small grains. Weeds also may harbor insect pests that normally are not important pests of small grains, but build up on weeds in small grain fields and attack subsequent susceptible crops in the same field or disperse to susceptible crops in adjacent fields. Wild radish and mustard weeds in winter wheat can harbor large numbers of lygus bugs and flower thrips (*Frankliniella* spp.) that disperse to adjacent, susceptible summer crops such as vegetables and cotton in the southern United States.

Volunteer small grains also can play a major role in serving as green bridging hosts for small grain pests. In the southern United States, the entire first generation of the Hessian fly occurs on volunteer wheat before wheat is normally sown in late fall. Furthermore, crop rotation can effectively reduce populations of insects with limited host ranges including Hessian fly, wheat strawworm, wheat jointworm, and wheat stem sawfly. The presence of volunteer wheat during fallow periods or in rotated crops may maintain a pest population thereby negating the benefit of crop rotation.

Weed–insect–crop interactions have not been studied extensively in North America. A better understanding of these potential interactions may provide insight to the control of arthropod pests through weed management.

Selected References. 6, 39

By G. David Buntin

Identification of Arthropods and Diagnosis of Injury

By G. David Buntin and Keith S. Pike

Effective management of small grain insect pests begins with accurate identification of the pests and reliable diagnosis of pest injury. Identification and diagnosis can be based upon three types of evidence: morphology, an insect's form and structure, presence: symptoms of an insect's injury, and situation: the time of occurrence during the season or location of a pest or its injury in a field or on a plant. A combination of the three types of evidence offers the greatest potential for accurate identification and diagnosis. Descriptions and color photographs of insects and the injury they cause and details about their time of occurrence are provided in the sections on individual insects in this handbook. However, the list of pests by injury type, the illustrations, and the keys in this section offer a quick method of identifying insects and diagnosing injury in small grains.

If the type of injury found in a small grain field can be categorized (e.g., feeding on leaves or tunneling in stalks), the list of small grain insects by injury type might provide a first clue about which insects are responsible. The page number after an insect's common name indicates the section in which the insect is discussed in detail.

The keys for diagnosis of injury and insect identification were designed for use in a stepwise fashion. At each step, two or three choices are presented, one of which should apply to the injury or insect to be identified. The choice made leads to another set of two or three choices, and so on until the name of the insect appears with an illustration. If the illustration does not match the injury or insect, users should retrace their steps to determine where the wrong choice was made. After the injury or insect has been identified with the keys, the color photographs can be studied to verify the diagnosis or identification.

The key for diagnosis of insect injury to small grains is presented as a series of choices. Different pests occur at different times of the season, so initial separators are times of the growing season. Other primary separators are the parts of the plants injured. When several insects cause the same type of injury at the same stage of small grain development, all of them are listed.

A positive association between the injury to the small grain plants and its cause is essential for making accurate pest management decisions. In some cases, injury alone will not distinguish which pest is causing the injury. Every attempt should be made to collect and identify the pest(s). Many species of insects found in small grains do not appear in the keys because their occurrence is incidental. The keys were designed for identifying the most common and important pest insects, and a few unusual small grain pests that often are mistaken for more serious pests.

The identification key of insects initially separates pests at or below the soil surface from pests that feed aboveground. Other major characteristics used to separate types of insects include the presence or absence, and types of legs and other structures; body shape and general size; types of wings; and color patterns.

The Glossary provides definitions of these terms.

Small Grain Arthropod Pests by Injury Type

Seed and Seedling Injury: Stand reduction
Army cutworm (p. 53)
Billbug (p. 49)
Fall armyworm (p. 46)
False wireworms (p. 85)
Lesser cornstalk borer (p. 64)
Pale western cutworm (p. 54)
Seedcorn maggot (p. 67)
White grubs (p. 83)
Wireworms (p. 84)

Root Feeding
Billbug larvae (p. 49)
False wireworms (p. 85)
Flea beetle (larvae) (p. 55)
Rice root aphid (p. 45)
White grubs (p. 83)
Wireworms (p. 84)

Leaf Feeding: Defoliation
Armyworm (p. 46)
Army cutworm (p. 53)
Blister beetles (p. 49)
Cereal leaf beetle (p. 50)
European corn borer (small larvae) (p. 68)
Fall armyworm (p. 46)
Flea beetle (p. 55)
Grasshoppers (p. 56)
Leaf sawfly (p. 62)
Mormon cricket (p. 65)
Pale western cutworm (p. 54)
Wheat head armyworm (p. 47)
Yellowstriped armyworm (p. 48)

Leaf Feeding: Leaf Chlorosis, Discoloration, and Distortion
Aphids (p. 37)
Banks grass mite (p. 78)
Brown wheat mite (p. 77)
Chinch bug (p. 52)
Leafhoppers (p. 63)
Plant bugs (p. 66)
Thrips (p. 72)
Wheat curl mite (p. 75)
Winter grain mite (p. 79)

Leaf Feeding: Leaf Mining
Grass sheathminer (p. 63)
Frit fly (p. 56)

Stem Boring
European corn borer (p. 68)
Frit fly (p. 56)
Hessian fly (p. 58)
Stalk borer (p. 69)
Wheat stem maggot (p. 79)
Wheat stem sawfly (p. 80)
Wheat strawworm (p. 82)
Wheat jointworm (p. 73)

Feeding on Grain Spikes and Kernels
Armyworm (large populations) (p. 46)
English grain and other Aphids (p. 39)
Plant bugs (p. 66)
Stink bugs (p. 70)
Thrips (p. 72)
Wheat curl mite (p. 75)
Wheat head armyworm (p. 47)
Wheat midge (p. 73)

Transmission of Plant Disease
Aphids (p. 37)
Brown wheat mite (p. 77)
Cereal leaf beetle (p. 50)
Leafhoppers (p. 63)
Planthoppers (p. 63)
Wheat curl mite (p. 75)

Small Grain Arthropod Pests by Scientific Classification

Phylum Arthropoda
  Class Arachnida
  Order Acari (Mites and Ticks)
  Family Eriophyidae
      Wheat curl mite (p. 75)
  Family Penthaleidae
      Winter grain mite (p. 79)
  Family Tetranychidae
      Banks grass mite (p. 78)
      Brown wheat mite (p. 77)

Class Insecta
  Order Orthoptera
  Family Acrididae
      Grasshoppers (p. 56)
  Family Tettigoniidae
      Mormon cricket (p. 65)

  Order Thysanoptera
  Family Thripidae
      Cereal Thrips (p. 72)
      Barley Thrips (p. 72)

  Order Hemiptera
  Suborder Heteroptera (True bugs)
  Family Lygaeidae
      Chinch bug (p. 52)
  Family Miridae
      Plant bugs (p. 66)
      Black grass bug (p. 66)
  Family Pentatomidae
      Stink bugs (p. 70)
  Suborder Sternorrhyncha (formerly Homoptera in part)
  Family Aphididae
      Aphids (p. 37)
  Suborder Auchenorrhyncha (formerly Homoptera in part)
  Family Cicidellidae
      Leafhoppers (p. 63)
  Family Delphacidae
      Planthoppers (p. 63)

Order Coleoptera (Beetles)
Family Scarabaeidae
      White grubs (p. 83)
Family Elateridae
      Wireworms (p. 84)
Family Tenebrionidae
      False wireworms (p. 85)
Family Meloidae
      Blister beetles (p. 49)
Family Chrysomelidae
      Cereal leaf beetle (p. 50)
      Flea beetle (p. 55)
Family Curculionidae
      Billbug (p. 49)

Order Lepidoptera (Moths and butterflies)
Family Pyralidae
      Lesser cornstalk borer (p. 64)
      European corn borer (p. 68)
Family Noctuidae
      Armyworm (p. 46)
      Army cutworm (p. 53)
      Fall armyworm (p. 46)
      Yellowstriped armyworm (p. 48)
      Pale western cutworm (p. 54)
      Wheat head armyworm (p. 47)
      Stalk borer (p. 69)

Order Diptera (True flies)
Family Cecidomyiidae
      Hessian fly (p. 58)
      Wheat midge (p. 73)
Family Agromyzidae
      Grass sheathminer (p. 63)
      Wheat stem maggot (p. 79)
Family Ottitidae
      Frit fly (p. 56)
Family Anthomyiidae
      Seedcorn maggot (p. 67)

Order Hymenoptera (Sawflies, wasps, ants, and bees)
Family Tenthredinidae
      Leaf sawfly (p. 62)
Family Cephidae
      Wheat stem sawfly (p. 80)
Family Eurytomidae
      Wheat jointworm (p. 73)
      Wheat strawworm (p. 82)

Key to Insect Injury to Wheat

By Tom Royer

Early Season Injury:
Planting to Stem Elongation

Start Here

Irregular germination; skips in drill rows where plants did not emerge
- **Seed corn maggot**
- **False wireworm**
- **Leatherjacket**
- **Wireworm**

Germination normal; plants injured after emergence

Leaf tissue removed

Plants cut at or just below the soil surface;
- **Army cutworm**
- **Pale western cutworm**
- **Fall armyworm**

Leaf tissue removed, chewed or torn, but plants not cut at soil surface;
- **Armyworm**
- **Beet armyworm**
- **Fall armyworm**
- **Flea beetle**
- **Grasshopper**
- **Leaf-feeding sawfly**
- **Yellowstriped armyworm**

Leaf tissue not removed

Plants stunted, or wilted, but primarily green in color, not dead

Leaf tissue discolored (red, yellow, white, brown, or portions of plants dead)

Plant tissue dead (necrotic)

At least some plant tissue discolored (purple, red, orange, yellow, white)
- **Chinch bug**
- **Greenbug**
- **Leafhopper**
- **Russian wheat aphid**
- **Yellow sugarcane aphid**

Green tissue removed, but white, transparent tissue remains, creating "windowpane" appearance
- **Cereal leaf beetle**
- **Flea beetle**
- **Leafminer**
- **Sheathminer**

Portions of plant crown dead, causing "dead heart"
- **Billbug**
- **Frit fly**
- **Lesser cornstalk borer**

Plant tissue stippled with bronze or silvery colored dead tissue.
- **Mites**
- **Thrips**

Leaves wilted because of root injury
- **False wireworm**
- **Lesser cornstalk borer**
- **White grub**
- **Wireworm**

Leaves thickened, stunted, and dark green to bluish green
- **Hessian fly**

Leaves wilted or slow growing, but not dark green
- **Bird cherry oat aphid**
- **Corn leaf aphid**

Key to Small Grain Pest Injury

Mid- to Late Season Injury: Stem Elongation to Maturity

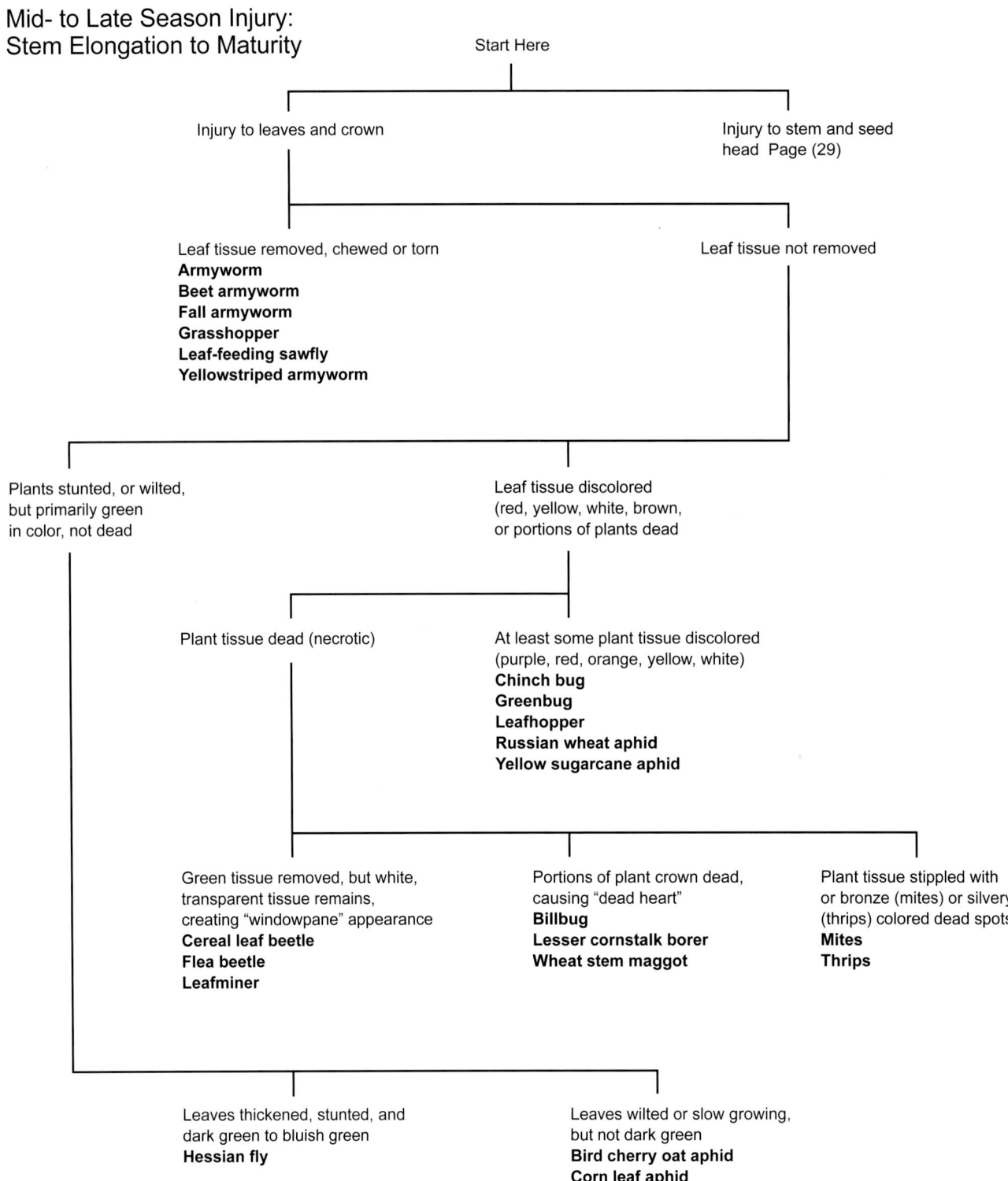

Key to Small Grain Pest Injury

Continued from page 28:

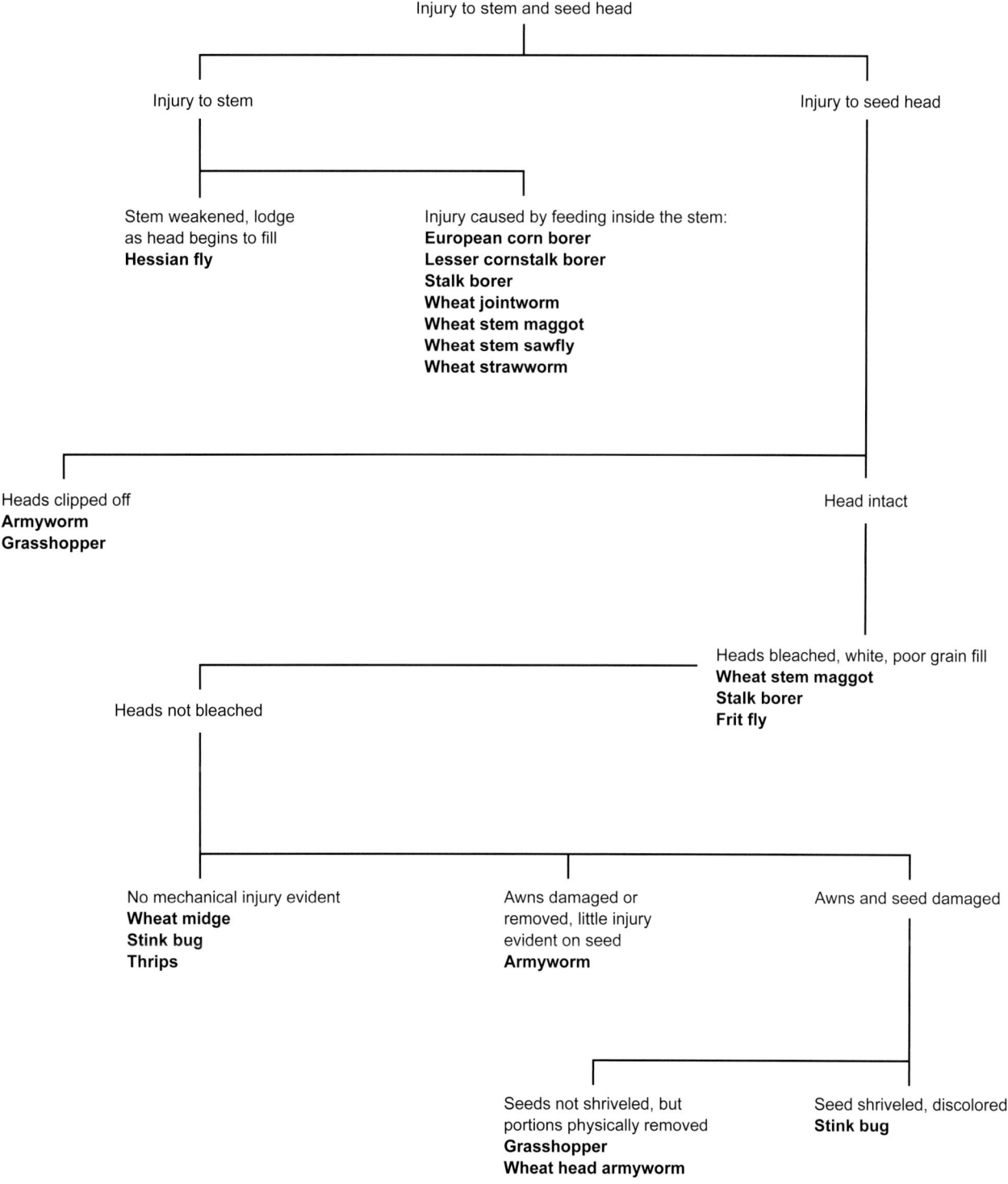

29

Key to Insect and Mite Pests of Small Grains

By Tom Royer

Wheat Insect Pests: Pest Activity Mostly Above the Soil Surface

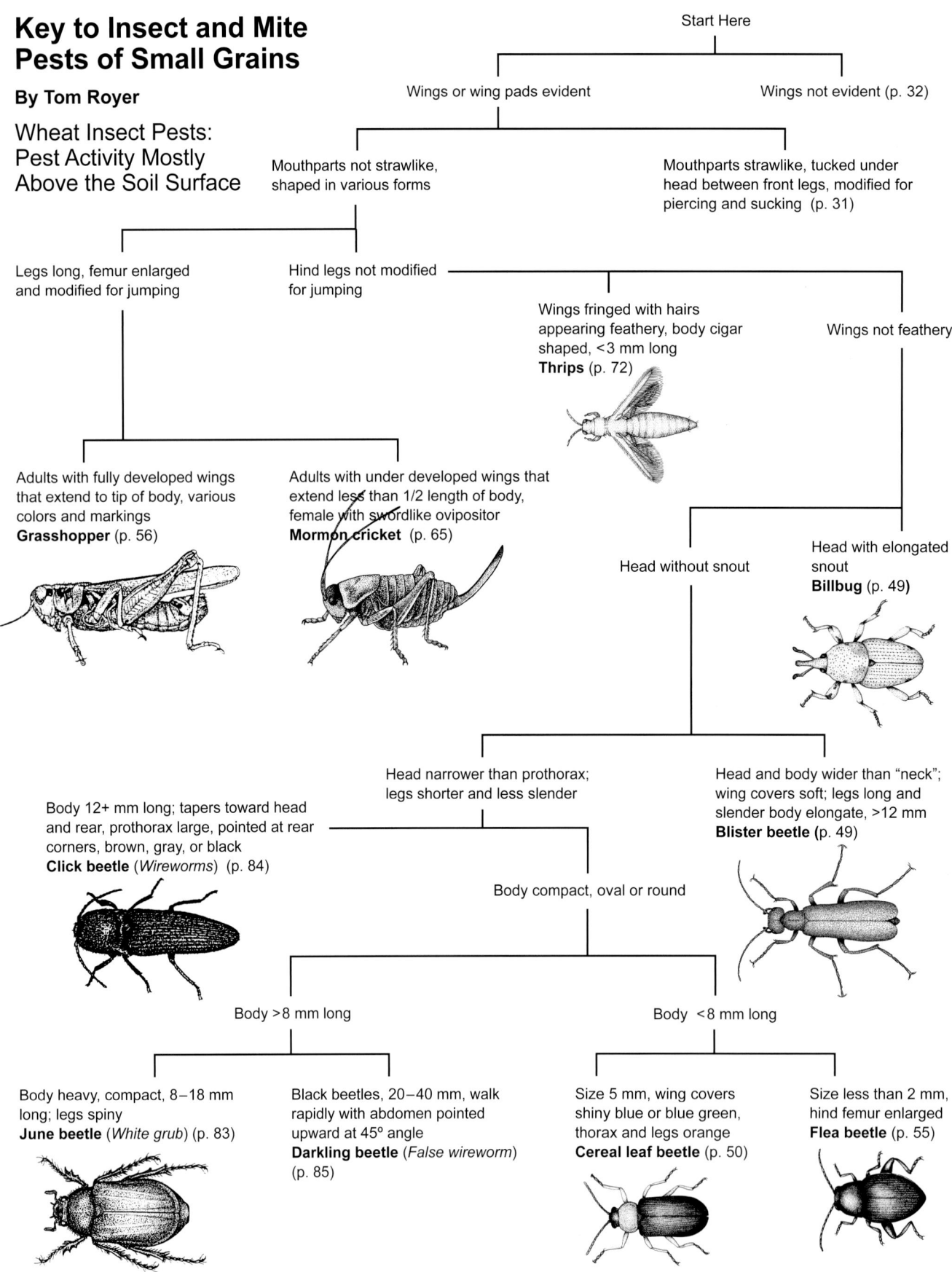

Key to Small Grain Insect Pests

Continued from page 30:

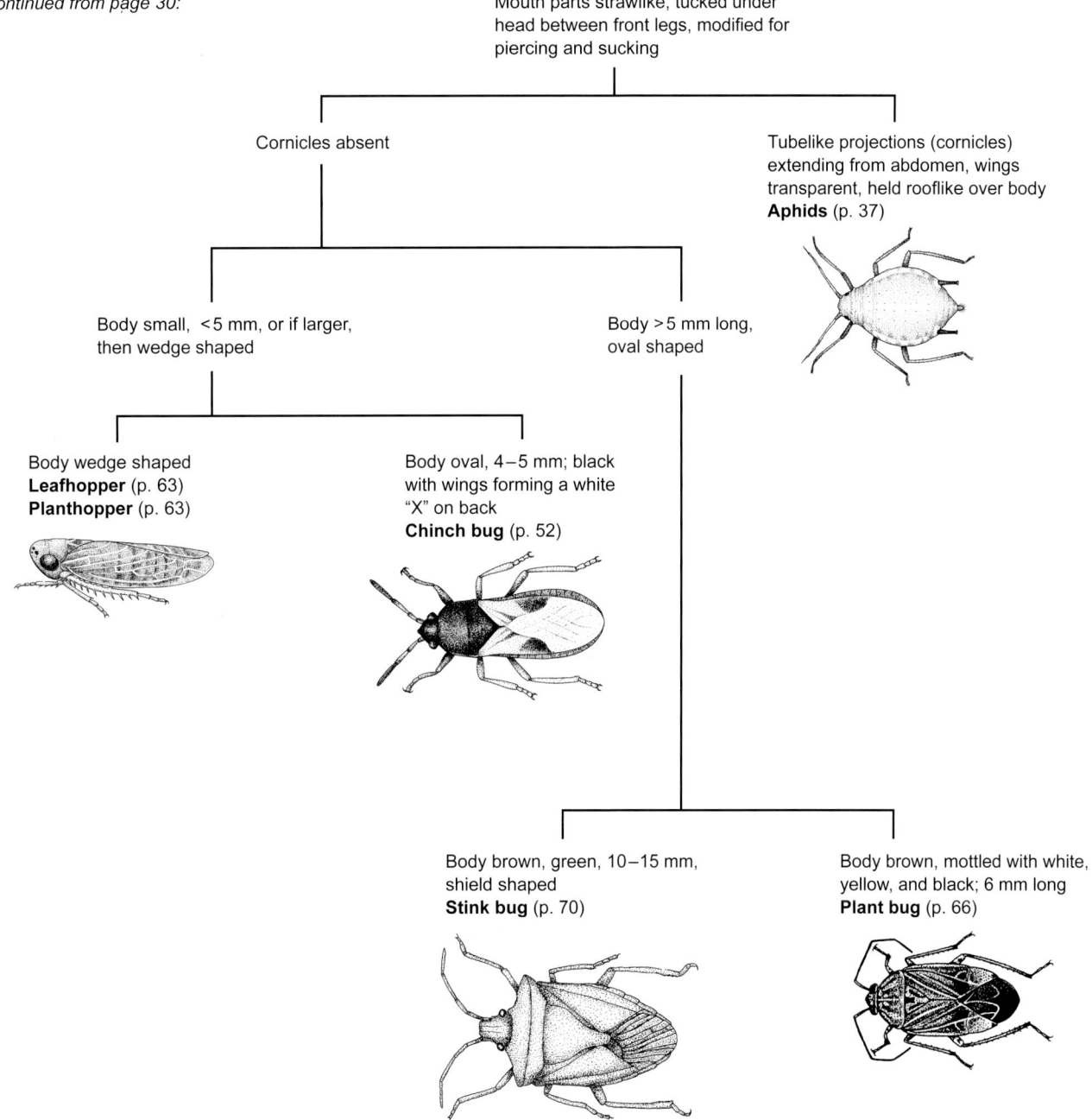

- Mouth parts strawlike, tucked under head between front legs, modified for piercing and sucking
  - Cornicles absent
    - Body small, <5 mm, or if larger, then wedge shaped
      - Body wedge shaped
        Leafhopper (p. 63)
        Planthopper (p. 63)
      - Body oval, 4–5 mm; black with wings forming a white "X" on back
        Chinch bug (p. 52)
    - Body >5 mm long, oval shaped
      - Body brown, green, 10–15 mm, shield shaped
        Stink bug (p. 70)
      - Body brown, mottled with white, yellow, and black; 6 mm long
        Plant bug (p. 66)
  - Tubelike projections (cornicles) extending from abdomen, wings transparent, held rooflike over body
    Aphids (p. 37)

Key to Small Grain Insect Pests

Continued from page 30:

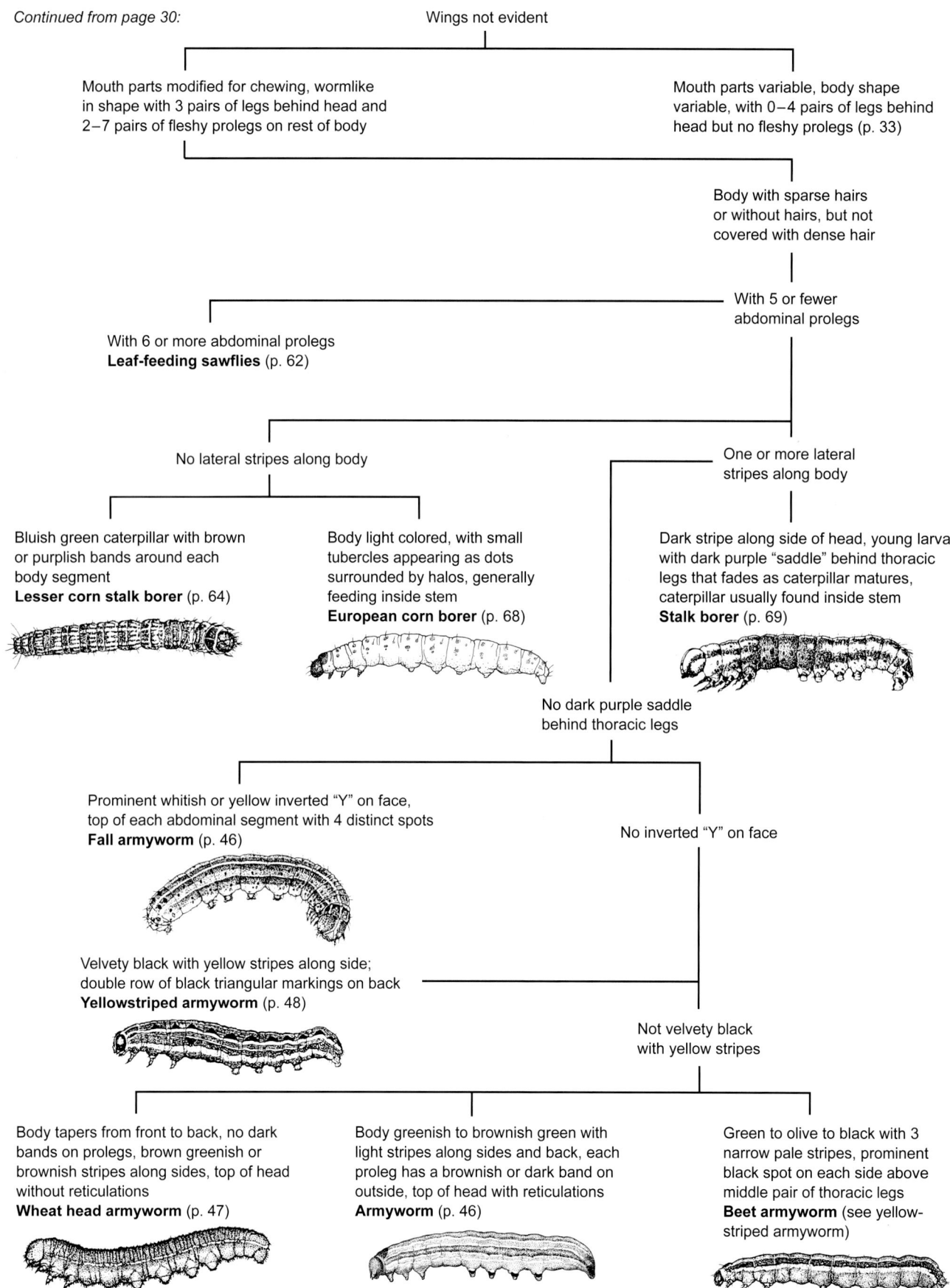

Wings not evident

- Mouth parts modified for chewing, wormlike in shape with 3 pairs of legs behind head and 2–7 pairs of fleshy prolegs on rest of body
- Mouth parts variable, body shape variable, with 0–4 pairs of legs behind head but no fleshy prolegs (p. 33)

Body with sparse hairs or without hairs, but not covered with dense hair

- With 6 or more abdominal prolegs
  Leaf-feeding sawflies (p. 62)
- With 5 or fewer abdominal prolegs

No lateral stripes along body

- Bluish green caterpillar with brown or purplish bands around each body segment
  Lesser corn stalk borer (p. 64)
- Body light colored, with small tubercles appearing as dots surrounded by halos, generally feeding inside stem
  European corn borer (p. 68)

One or more lateral stripes along body

- Dark stripe along side of head, young larvae with dark purple "saddle" behind thoracic legs that fades as caterpillar matures, caterpillar usually found inside stem
  Stalk borer (p. 69)

No dark purple saddle behind thoracic legs

- Prominent whitish or yellow inverted "Y" on face, top of each abdominal segment with 4 distinct spots
  Fall armyworm (p. 46)
- No inverted "Y" on face

- Velvety black with yellow stripes along side; double row of black triangular markings on back
  Yellowstriped armyworm (p. 48)
- Not velvety black with yellow stripes

- Body tapers from front to back, no dark bands on prolegs, brown greenish or brownish stripes along sides, top of head without reticulations
  Wheat head armyworm (p. 47)
- Body greenish to brownish green with light stripes along sides and back, each proleg has a brownish or dark band on outside, top of head with reticulations
  Armyworm (p. 46)
- Green to olive to black with 3 narrow pale stripes, prominent black spot on each side above middle pair of thoracic legs
  Beet armyworm (see yellow-striped armyworm)

Key to Small Grain Insect Pests

Continued from page 32:

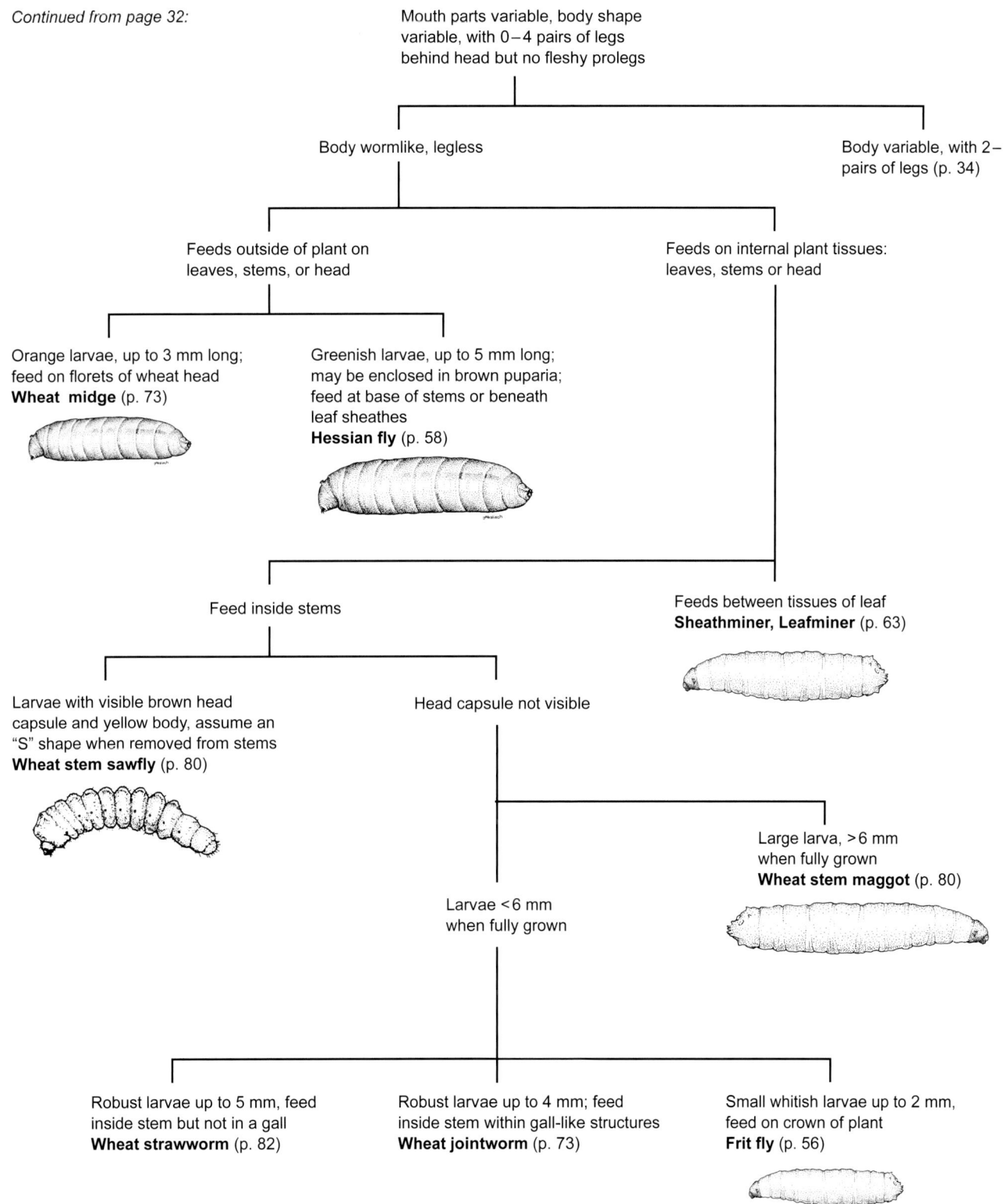

Mouth parts variable, body shape variable, with 0–4 pairs of legs behind head but no fleshy prolegs

- Body wormlike, legless
- Body variable, with 2–4 pairs of legs (p. 34)

Body wormlike, legless:
- Feeds outside of plant on leaves, stems, or head
- Feeds on internal plant tissues: leaves, stems or head

Feeds outside of plant on leaves, stems, or head:
- Orange larvae, up to 3 mm long; feed on florets of wheat head
  Wheat midge (p. 73)
- Greenish larvae, up to 5 mm long; may be enclosed in brown puparia; feed at base of stems or beneath leaf sheathes
  Hessian fly (p. 58)

Feeds on internal plant tissues:
- Feed inside stems
- Feeds between tissues of leaf
  Sheathminer, Leafminer (p. 63)

Feed inside stems:
- Larvae with visible brown head capsule and yellow body, assume an "S" shape when removed from stems
  Wheat stem sawfly (p. 80)
- Head capsule not visible

Head capsule not visible:
- Larvae <6 mm when fully grown
- Large larva, >6 mm when fully grown
  Wheat stem maggot (p. 80)

Larvae <6 mm when fully grown:
- Robust larvae up to 5 mm, feed inside stem but not in a gall
  Wheat strawworm (p. 82)
- Robust larvae up to 4 mm; feed inside stem within gall-like structures
  Wheat jointworm (p. 73)
- Small whitish larvae up to 2 mm, feed on crown of plant
  Frit fly (p. 56)

Key to Small Grain Insect Pests

Continued from page 33:

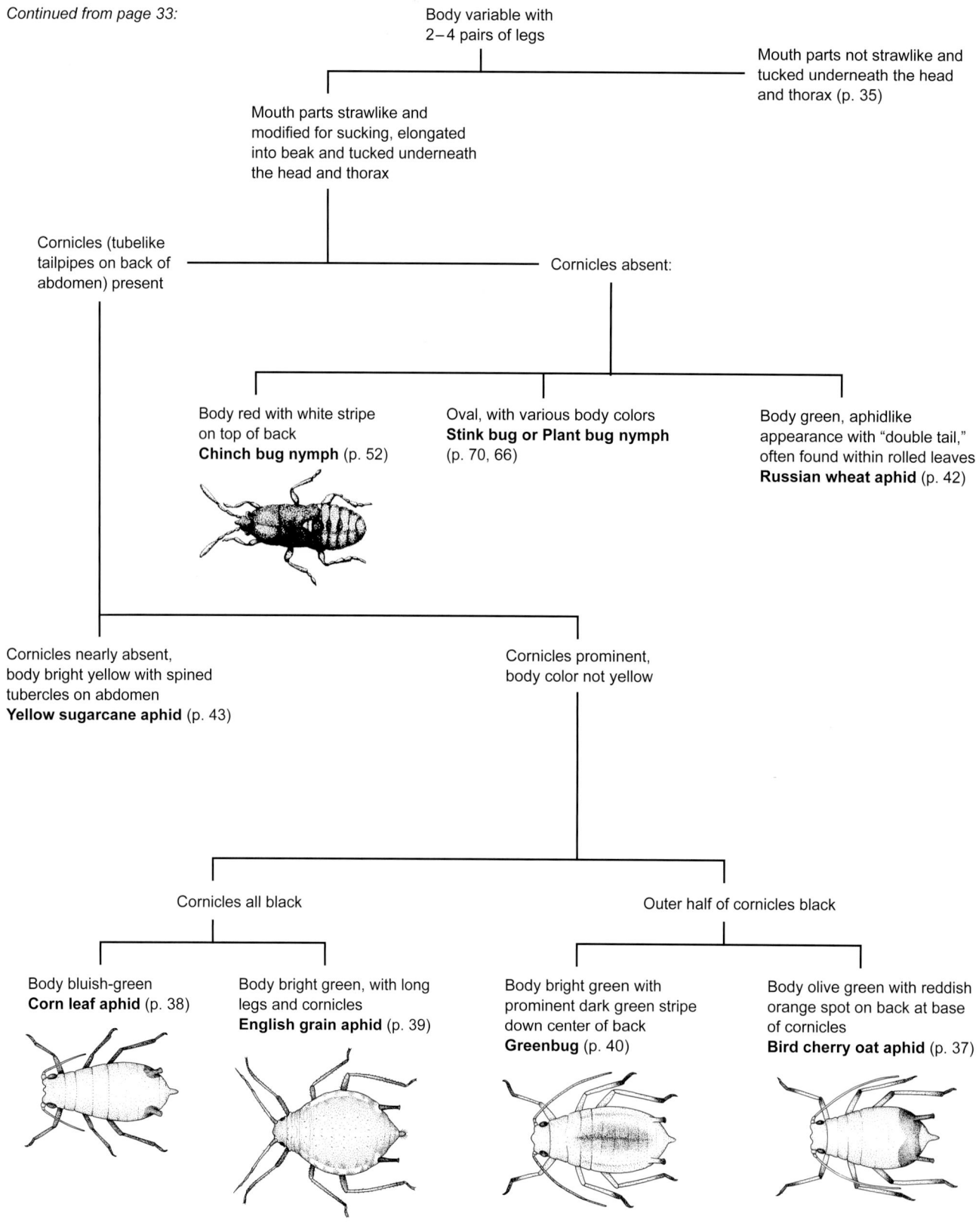

- Body variable with 2–4 pairs of legs
  - Mouth parts strawlike and modified for sucking, elongated into beak and tucked underneath the head and thorax
    - Cornicles (tubelike tailpipes on back of abdomen) present
      - Cornicles nearly absent, body bright yellow with spined tubercles on abdomen
        Yellow sugarcane aphid (p. 43)
      - Cornicles prominent, body color not yellow
        - Cornicles all black
          - Body bluish-green
            Corn leaf aphid (p. 38)
          - Body bright green, with long legs and cornicles
            English grain aphid (p. 39)
        - Outer half of cornicles black
          - Body bright green with prominent dark green stripe down center of back
            Greenbug (p. 40)
          - Body olive green with reddish orange spot on back at base of cornicles
            Bird cherry oat aphid (p. 37)
    - Cornicles absent:
      - Body red with white stripe on top of back
        Chinch bug nymph (p. 52)
      - Oval, with various body colors
        Stink bug or Plant bug nymph (p. 70, 66)
      - Body green, aphidlike appearance with "double tail," often found within rolled leaves
        Russian wheat aphid (p. 42)
  - Mouth parts not strawlike and tucked underneath the head and thorax (p. 35)

Key to Small Grain Insect Pests

Continued from page 34:

Mouthparts not strawlike and tucked underneath the head and thorax

Body carrot shaped, microscopic, with two pair of legs present just behind head
Wheat curl mite (p. 75)

Body oval, with 4 pair of legs (3 pair on larvae) spiderlike in appearance

Front pair of legs nearly twice as long as other three pair, body brown
Brown wheat mite (p. 77)

Front pair of legs about same length as other three

Body with orange legs and an orange spot on top of abdomen, feed at base of plant
Winter grain mite (p. 79)

Dark areas along side of body forming line, extending to rear, feed on leaf blades
Banks grass mite (p. 78)

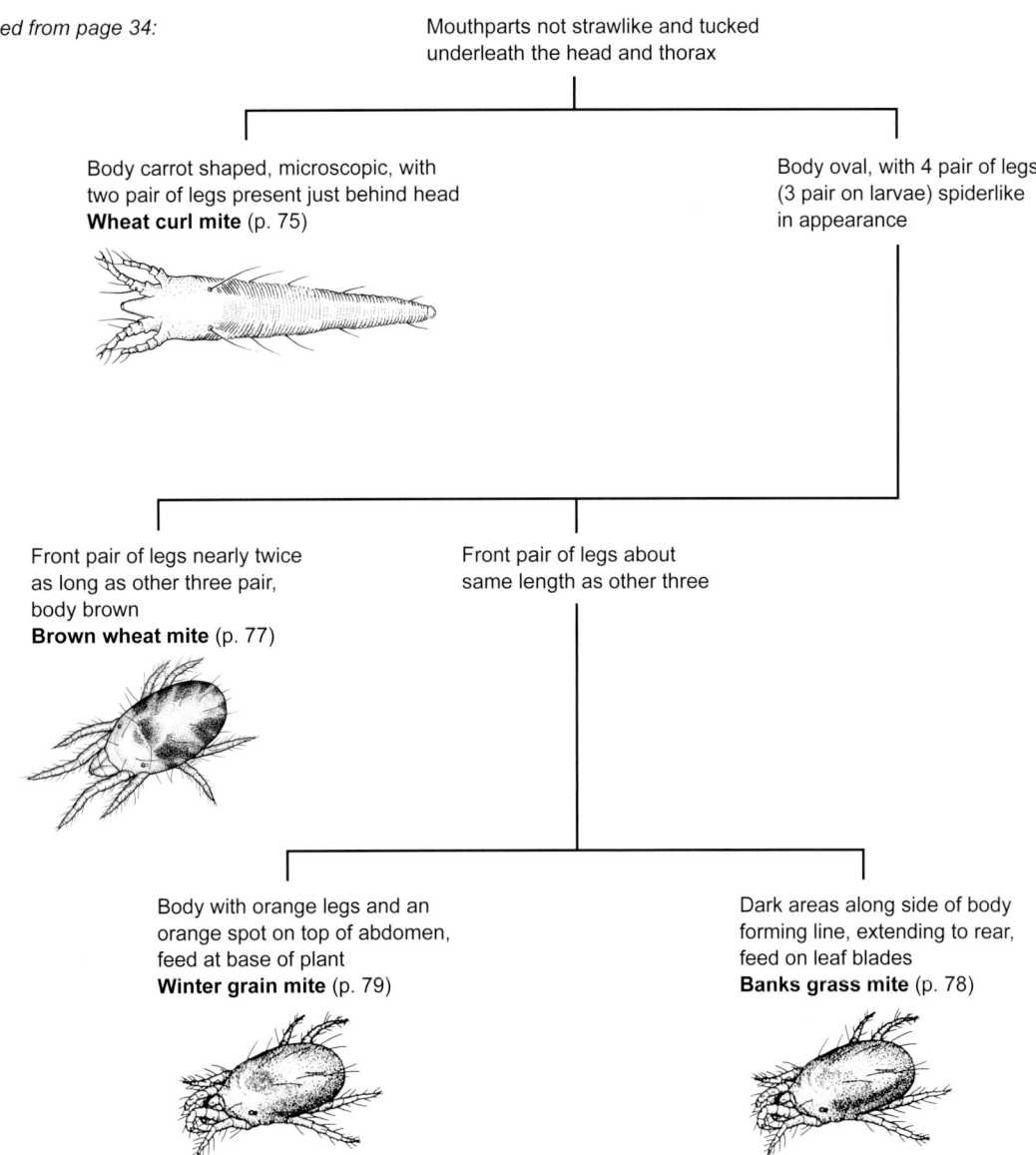

Key to Small Grain Insect Pests

Wheat Insect Pests: Pest Activity Mostly Below the Soil Surface

Start Here

- **Legs present**
  - **More than 3 pairs of legs present, 3 just behind the head, plus 3–5 pairs of fleshy prolegs**
    - Fleshy caterpillar, striped or plain, curls into a tight "C" when disturbed
      Cutworm (p. 53)
  - **Three pairs of legs**
    - Body "C" shaped, with brown head and long brownish legs, dark area on swollen tail segment
      White grub (p. 83)
    - Body long, cylindrical, yellowish to brown, legs short
      Wireworm or **False wireworm** (p. 84)
    - Bluish green caterpillar with brown or purplish bands around each body segment
      Lesser cornstalk borer (p. 64)

- **Legs absent**
  - Head absent or not visible
    - Body yellowish white, 6 mm or less in length, "maggotlike" shape
      Seed corn maggot (p. 67)
  - Head capsule visible, tan, body smooth and swollen
    Billbug larva (p. 49)

Pest Information

The following information is provided for each pest. The information is presented in a standardized format within the subheadings listed here. More detailed information is provided for the most important small grains pests in part because information on minor pests is limited or not available.

Common Name: Common name of the pest is listed at the beginning of the chapter. If the Entomological Society of America has designated an approved common name for a species, this name is used. If a pest is widely known by other common names, they may be listed.

Scientific Classification. The genus and species are followed by order and family names (family name is in parentheses). Recent or widely cited former scientific names also are listed.

Origin and Distribution. This section indicates whether the pest is native or exotic and gives its approximate distribution in North America

Description. Characteristic identification details are given for injurious stages and sometimes for noninjurious stages.

Pest Status. This section describes the economic importance of the pest throughout its range.

Injury. This section describes how the pest injures small grain plants.

Life History. Details about the pest's life cycle are provided, such as information on overwintering stage(s), number of generations per year, developmental times, location of each stage, and time of occurrence. Details of a pest's life history represent average or typical values, which may vary by location and climatic conditions.

Management. General management tactics are described. When known, information on the effect of biological controls is discussed. Widely used economic thresholds and other decision-making tools are listed. Insecticidal control is discussed, but specific insecticides are not given.

Selected References. A number of general and specific sources of additional information are listed.

Distribution Map. The distribution of the pest in North America is indicated on the map. A distribution map is not included if the pest occurs throughout North America wherever small grains are grown, or if pest's distribution could not be accurately determined. Distributions are approximations, and the pest may occur outside the range indicated. In particular, demarcation of the northern and southern limits of their range may be uncertain.

Photographs. Color photographs of the injurious stage and/or injury caused by the pest are included for most species.

Aphids

At least 20 species of aphids have been reported to infest small grains in North America. The six most important species are presented separately because each species poses a unique threat to small grain production. Four other minor species are discussed under miscellaneous aphids. Aphids are key pests of small grains throughout North America in part because they transmit viral pathogens. The most important disease transmitted by aphids is barley and cereal yellow dwarf, which is briefly discussed on page 20.

Bird Cherry-Oat Aphid

Scientific Classification. *Rhopalosiphum padi* (L.). Hemiptera: Sternorrhyncha (Aphididae).

Origin and Distribution. The aphid has long been considered Palearctic in origin, but recent findings suggest that it is Nearctic. Today, it is worldwide in its distribution.

Description. Adult measures 1/32–1/16 in. (0.8–1.6 mm) long. Body is usually dark green, but can vary from green to black; antennae and tarsi typically dark; legs green to pale green; siphunculi green to dark green with black apices; abdomen with a reddish-orange patch between and at the base of the siphunculi. In winged forms (cleared and mounted), abdomen pale with marginal pigmentation, spots and/or flecks; siphunculi constricted before the flange, indistinctly swollen subapically; antennae 6-segmented with rhinaria on segments III–V, and terminal process 3-6x base of VI; forewing with media usually twice-branched (second branch short, sometimes nonexistent).

Pest Information

Bird cherry oat aphid (Keith S. Pike).

Pest Status. The aphid is the premier vector of barley yellow dwarf virus (BYDV), and often the most abundant cereal aphid species on winter grains in the fall and winter. It also occurs commonly in high numbers in mid- to late summer on corn in northwestern United States.

Injury. It feeds on all types of small grains. Primary crop damage is caused by BYDV transmitted by the aphid, resulting in stunted plants, small heads, shriveled kernels, and reduced yields. Direct feeding, though less important, can also adversely affect crop health, especially if a crop is infested during early seedling stages.

Life History. The life cycle involves sexual and parthenogenic viviparous reproductive phases. The aphid migrates annually (heteroecious holocyclic) between its overwintering hosts, chokecherry, *Prunus virginiana* or bird cherry, *P. padus* (latter species is European, known only as an introduced ornamental in North America) and its summer hosts, Gramineae (grasses) in northern United States and Canada. In areas with warm winters, it stays on Gramineae all year (anholocyclic). In northern areas especially, the aphid responds to decreasing photoperiod and temperature in the fall to produce winged migrants (gynoparae and males) that move from Gramineae to *Prunus* spp. Oviparae, produced by gynoparae and fertilized by males, deposit eggs on the primary host from September through October. Eggs hatch around mid-April (timing varies with altitude and latitude). The resulting apterae are green, wax-covered and feed on the underside of young leaves. Subsequent spring migrants return to suitable Gramineae summer hosts between May and July. Summer generations comprise parthenogenic viviparae, and varying proportions of these remain as viviparae (anholocyclic forms) through the fall and winter on Gramineae, most successfully and predominantly in areas of mild winters.

Management. The aphid is relatively easy to control with chemicals, but economic injury thresholds and management measures vary between states and regions because of differences in aphid dynamics, virus transmission, cropping schemes, economics, and natural enemies. Thus, it is best to consult local Cooperative Extension personnel or other appropriate sources for thresholds and control recommendations. Populations peak between October and March, depending upon latitude. Survival of anholocyclic forms in northern areas is typically very low in the spring. The aphid is attacked by a variety of parasitoids, predators, and entomopathogenic fungi, which play a significant role in maintaining the aphid at generally low levels.

Selected References. 32, 36, 144, 155

By Keith S. Pike

Corn Leaf Aphid

Scientific Classification. *Rhopalosiphum maidis* Fitch. Hemiptera: Sternorrhyncha (Aphididae).

Origin and Distribution. The corn leaf aphid is native to North America. It survives year-round in the Gulf Coast states and migrates north throughout the United States and into Canada each season.

Description. The adult corn leaf aphid is between 1/16 and 1/12 in. (1.8–2.1 mm) long. This blue-green aphid has black legs, antennae, eyes, and short broad siphunculi (sometimes called tailpipes) with a dark spot on the abdomen at the base of the siphunculi. Adults may become almost completely dark green to black as they age.

Corn leaf aphid (Keith S. Pike).

Pest Status. Insecticides applied for control of corn leaf aphid have rarely resulted in an increase in grain quantity or quality. Although not a direct pest, corn leaf aphid is a vector of barley and cereal yellow dwarf viruses, and its control may be important.

Injury. Like other aphids attacking grain crops, this species causes damage by removing plant sap with piercing sucking mouthparts and by disseminating disease-producing virus. Corn leaf aphid can be found on all types of small grains but is more common on wheat and barley.

Life History. The life cycle of this species is similar to that of other aphid species. Corn leaf aphid is anholocyclic producing live young by parthenogenesis. It feeds on numerous grass species and has several nymphal instars. Populations become established in northern areas by migration from the south. Numerous generations may occur every year; the exact number affected by the latitude of the location.

Management. In most cases, producers do not use management techniques to control corn leaf aphid on small grains. No specific treatment thresholds are available for corn leaf aphids. Some states recommend treatment for aphids at various thresholds. In some areas, delaying planting or avoiding early planting lowers the chance of aphid activity and barley yellow dwarf infection.

Selected References. 44

By Gerald E. Wilde

English grain aphids (Jay W. Chapin).

Feeding injury by English grain aphid (Jay W. Chapin).

English Grain Aphid

Scientific Classification. *Sitobion avenae* (F.). Hemiptera: Sternorrhyncha (Aphididae).

Origin and Distribution. English grain aphid is found wherever cereals are produced, except Australia. An introduced Old World species, it now occurs throughout North America, predominantly on cereal crops, but also on corn and grasses such as foxtail grasses, *Setaria* spp.

Description. Wingless forms are up to 1/8 in. (3.2 mm), winged forms ~1/5 in. (5.1 mm) long. Color of wingless forms is variable, ranging from green to yellow, pink, red, or even dark brown. Some color forms appear to be associated with particular plants, such as green ones on corn, but all colors can be found on small grains. Siphunculi projecting posteriorly from the abdomen are entirely black except in small nymphs. The cauda is pale. Eyes are typically red. The dark antennae extend straight back and are over half the length of the body. Legs are dark or have dark bands near leg joints. Small nymphs may have a green line down the middle of the back, similar to the greenbug. Winged forms have distinct black siphunculi, light and dark banded legs, and a pale cauda.

Pest Status. Although widely distributed, English grain aphid is not regarded as a significant economic pest in most of North America. However, it is the most important cereal aphid in Manitoba and Saskatchewan. In the Pacific Northwest and northern Great Plains, English grain aphid is a sporadic pest on spring wheat. During outbreak years in the southeastern United States, English grain aphid can cause damage across a multistate area. Economic injury has been documented on wheat, oats, barley, and rye.

Injury. Direct injury is caused by the removal of carbohydrates and nitrogen as the aphid extracts plant sap from the upper leaves, particularly the flag leaf, and grain heads. Once the green head emerges, English grain aphids are typically found feeding head down at the base of glumes. This feeding can reduce seed number, seed weight, grain protein level, and baking quality of flour. The greatest potential for injury occurs during flowering and early grain formation. After flowering, seed number is not affected. Indirect loss may occur from a reduced photosynthesis caused by honeydew and increases in saprophytic fungi associated with aphid feeding sites. In the southeastern United States, feeding colonies can cause chlorotic or necrotic lesions on flag leaves. As the crop approaches maturity, affected fields may exhibit dark "hits", corresponding to areas of high aphid infestation. In these areas, the flag leaves have senesced prematurely, and the heads are darkened by secondary fungal infection. English grain aphid feeding has been associated with symptoms of "black chaff" disease, which is caused by the bacterium *Xanthomonas campestris*. English grain aphid also can reduce shoot and root growth of cereals when seedlings are infested. Yield loss from seedling infestation has been documented on spring and winter wheat. English grain aphid also is a vector of BYDV, primarily the MAV and PAV strains. However, in most areas of North America, English grain aphid plays a secondary role in virus transmission.

Life History. The life cycle is relatively simple. In the summer, all aphids are females that give birth to live young. In northern climates, in response to cooler temperatures and shorter days, colonies produce egg-laying females and the only males of the season. These mate to produce winter-hardy eggs, laid predominantly in winter wheat. Eggs hatch into live-bearing females in early spring. English grain aphid does not overwinter in the central Canadian plains or in the northern Great Plains. In warmer latitudes, the sexual phase of the life cycle may not occur. Summer is spent on corn and other grasses, and winter on cereals. Live-bearing females can be either winged or wingless. When colonies become crowded, or host plant quality declines (as when cereals mature), the next generation of female aphids is winged. These disperse to colonize new host plants. At optimal temperatures (~25 °C), nymphs mature in about a week and begin producing the next generation of nymphs at a rate of 2–3/d for 10 d or more.

Natural enemies include numerous species of ladybird beetles and parasitic wasps. In drier climates, pathogenic fungi and syrphid fly larvae may not be as important in controlling English grain aphid as they are with other cereal aphids, because English grain aphid tends to live in exposed locations (e.g., heads of maturing cereals) where the humidity is low. Wind gusts and heavy rain can dislodge English grain aphid from grain heads and upper leaves causing high mortality in just a few hours.

Management. Treatment thresholds for English grain aphid vary considerably across North America, in part because of local differences in the frequency of outbreaks, yield expectations, and the potential for English grain aphid to multiply and feed on heads for an extended period. In the southeastern United States, growth-stage-dependent thresholds are used: 2–3 aphids per stem during stem elongation, 5 per stem at flag leaf to head emergence, 10 per stem from head emergence to milk stage, and too late to treat at soft dough stage. When sampling for English grain aphid, individual stem counts are more efficient than sampling a length of row or unit area. The usual recommendation is to count aphids on 5–10 stems from several areas of a field to estimate aphids per stem. Binomial sequential sampling plans are efficient because they only require an estimate of percentage of infested stems. For example, a 67% stem infestation level is almost equivalent to 5 aphids per stem. In practice, treatment decisions are often delayed until aphids become most noticeable at peak populations. Usually such treatments are not profitable because most of the injury has already occurred, and the aphid population will soon decline without treatment as the host plant matures. In areas where other pests, such as bird cherry-oat aphid or cereal leaf beetle, require treatment before English grain aphid attacks wheat, proper insecticide selection can eliminate the potential for English grain aphid damage by providing adequate residual control. Delayed planting can significantly reduce English grain aphid infestations on winter wheat, but this practice often is impractical because late-planted wheat has reduced yield potential.

Selected References. 15, 36, 90, 141, 196

By Jay W. Chapin and Susan E. Halbert

Greenbug

Scientific Classification. *Schizaphis graminum* (Rondani). Hemiptera: Sternorrhyncha (Aphididae).

Origin and Distribution. The earliest report of the greenbug was in Italy in 1847. It was first discovered in the United States in 1882; and severe damage was reported as far west as Texas, Missouri, and Indiana in 1890. The greenbug is reported from all continents except Australia. In the United States, it occurs in all areas where cereal grains or sorghum are grown.

Description. The greenbug is a small, oval-shaped, light-green aphid that usually has a dark green stripe down the back. Adults are 1/16–1/8 in. (1.6–2.1 mm) long. Two conspicuous siphunculi, often referred to as "tail-pipes", arise from near the end of the abdomen. Except for dark tips on the siphunculi and legs, these appendages are light colored. The size, shape, and color of the siphunculi are among the easiest ways to distinguish greenbugs from most other common aphids found on cereal grains.

Pest Status. More than 70 grass species have been identified as hosts for greenbugs, but serious damage is usually limited to wheat, oats, barley, sorghum, and bluegrass. The greenbug is one of the most important insect pests of wheat and sorghum in North America and is the most important insect pest of wheat and sorghum in portions of

Greenbugs (Jim Kalisch, University of Nebraska).

the High Plains states.

Injury. Greenbugs extract sap with their needlelike mouthparts and inject toxic saliva into plants. Initial feeding damage often appears as a yellowish spot with a small dark lesion in the center. With continued feeding, the leaves may turn yellow and die. From a distance, leaf damage may appear similar to symptoms of nitrogen deficiency or moisture stress. The highest concentration of greenbugs is usually found on the lower surface of lower leaves. Although greenbugs may occur throughout the field, infestations often appear as large, expanding yellowish circular areas. In addition to direct feeding damage, greenbugs vector plant pathogens such as BYDV.

Life History. Winged and wingless greenbugs may be found in wheat. All wingless greenbugs are female; and throughout most of the year, they give birth by parthenogenesis to living young, all of which are female. Although variable among biotypes and on different hosts, under optimal conditions (24–27 °C) greenbugs generally begin reproducing ~7 d after birth, and one female can produce 60–80 offspring over a 20–25-d period. In southern wheat-growing areas, greenbug adults and nymphs may be present in wheat throughout the winter and actively feed and reproduce on warm days. Additionally, in late fall, when there is <12 h of daylight, greenbug sexual morphs may be produced and lay eggs; greenbugs may overwinter in the egg stage. Eggs are probably a more important overwintering mechanism in northern wheat-producing areas where greenbug adults and nymphs cannot survive cold winter conditions.

Management. Greenbug-resistant varieties, biological control, and insecticides are all important components of a greenbug management program. Although greenbug-resistant varieties are often effective in reducing damage, greenbug populations have changed over time, and new biotypes (strains, designated by letters) became prevalent that were able to damage previously resistant varieties. From 1961 to 1995, 10 biotypes (A, B, C, E, F, G, H, I, J, and K) capable of damaging wheat were identified. For greenbug-resistant varieties to be effective in reducing greenbug damage, the wheat variety selected must be resistant (if available) to the predominant biotype in the area.

Several species of lady beetles, lacewings, syrphid fly larvae, damsel bugs, and several parasitoids are often effective in reducing greenbug damage. Greenbugs develop and reproduce at cooler temperatures than many of their predators and parasitoids, and greenbug damage is often more serious when spring conditions are cooler than normal. Under favorable conditions, 1–2 lady beetles per linear foot of row can control light-to-moderate infestations of greenbugs. Parasitoid eggs are laid, and larvae develop inside greenbugs, killing them in 4–7 d. The enlarged body of the dead greenbug, containing the parasitoid pupae, is golden brown or dark blue-black depending on the species of parasitoid; it is referred to as a "mummy." Once 10–20% of the greenbugs become mummies, greenbug populations are usually brought under control within 7–10 d. Under moist conditions, pathogenic fungi may be highly effective in reducing aphid numbers. Additionally, adverse climatic conditions, such as hard driving rains or hot dry winds, can reduce greenbug populations.

Because a variety of factors such as predators, parasitoids, pathogens, and adverse climatic conditions may reduce greenbug populations, the use of insecticides should be delayed until the economic threshold is reached. Generalized economic thresholds are listed here:

Plant Growth Stage	Economic Threshold Greenbugs/linear foot of row
Seedling	25–50
Plants 3–6 in. tall	100–300
Plants >6 in. tall	300–600

Greenbugs are seldom a problem after jointing.

Greenbugs resistant to many recommended insecticides, particularly organophosphate insecticides, are often present in low numbers in greenbug populations. Use of foliar applications of insecticides to control greenbugs or other insects should be delayed as long as possible to reduce the chance of insecticide-resistant greenbugs causing future problems. Seed treatments or soil insecticide applications at planting will control early season infestations of greenbugs and may be effective in areas that consistently have severe early season greenbug damage. For specific information on greenbugs and insecticide recommenda-

Pest Information

tions in various areas of the United States, consult local state extension entomologists.

Selected References. 11, 13, 62, 83, 163

By Z B Mayo

Russian Wheat Aphid

Scientific Classification. *Diuraphis noxia* (Kurdjumov). Hemiptera: Sternorrhyncha (Aphididae).

Origin and Distribution. Russian wheat aphid is native to eastern Europe and western Asia. It was first found in North America in 1980 near Mexico City and in the United States in 1986 near Muleshoe, TX. It is currently found in the wheat- and barley-producing areas of the 16 western states, with an eastern limit at about the 100th meridian.

Description. Russian wheat aphid is a small, 1/16–1/12 in. (1.6–2.1 mm) long, spindle-shaped, lime-green aphid, which is distinguished from other small grain aphids by having short antennae, having greatly reduced siphunculi, and appearing to have a double cauda (cauda and supracaudal process) when viewed from the side. Western wheat aphid, *D. tritici* (Gillette), is similar in its shape, size, and damage to wheat, but it has a single cauda and is much waxier in appearance.

Pest Status. Russian wheat aphid is a key pest of small grains in the western United States. Direct economic losses due to lower production and increased insecticide costs totaled ~$500 million from 1987 to 1994. Economic impact has varied from year to year and from location to location. Eastern Colorado has had relatively consistent infestations, accounting for ~30% of total insecticide use against this pest.

Injury. Russian wheat aphid feeds on younger leaves of wheat, barley, and other cool season grasses, preventing normal unrolling of leaves. Damaged leaves remain tightly rolled and have longitudinal white streaks and, less frequently, areas of purplish discoloration. Infestations may occur throughout wheat growth and development, but are less successful after head emergence. The severity of damage is related to length of infestations. Severe infestations can cause stunted plants and heads, bleached heads containing poorly formed grain, and poorly emerged heads caused by awns becoming trapped in the tightly rolled flag leaf. Infestations often result in reduced grain yield and quality. Russian wheat aphid is not known to transmit any important cereal diseases.

Life History. Russian wheat aphids remain on their hosts except during dispersal, which is accomplished primarily by the winged form. Important hosts are wheat and

Russian wheat aphids (Frank B. Peairs).

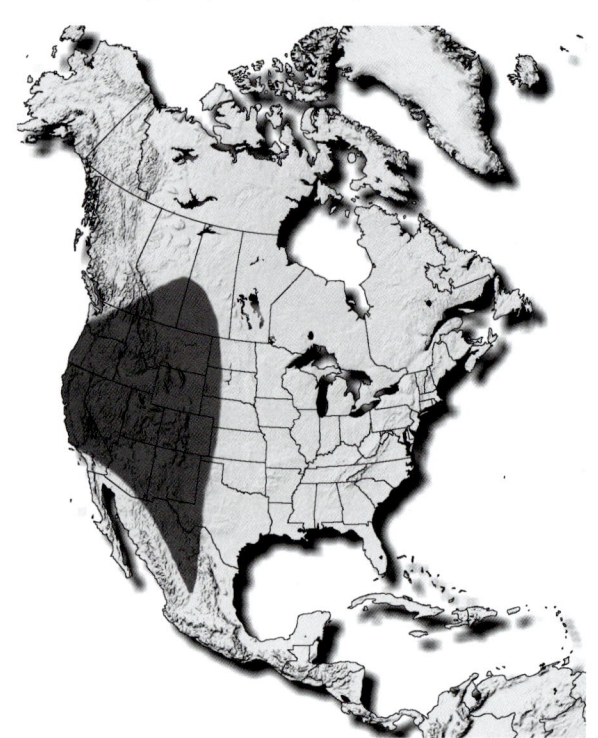

Russian wheat aphid distribution.

barley, and most of the year is spent on these crops. As the crop matures, these aphids will move to new hosts over short or long distances. These can be either less mature wheat and barley crops or a number of cool-season grasses, including crested wheatgrass, intermediate wheatgrass, and Canada wildrye.

Adult females give birth to live young, which mature and start to reproduce within 2 wk, or more, depending on temperature. Male Russian wheat aphids have not been found in North America. Individuals live for 60–80 d and produce ~80 offspring under favorable conditions. Approximately 285 degree days (DD) >39 °F (4 °C) are required from birth to reproduction. Optimum temperature for growth and development is ~68 °F (20 °C).

Russian wheat aphid is sufficiently cold tolerant to survive most winters in the Pacific Northwest and in the Great Plains as far north as the Colorado–Nebraska border. Individual aphids can survive temperatures as low as –13 °F (–25 °C), although exposure to temperatures <14 °F (–10 °C) reduces reproduction and life span. Continuous snow cover for >40 d also reduces overwintering success.

Biological control agents have been imported and released against Russian wheat aphid. In addition, many of the natural enemies of other cereal aphids (predators, parasitoids, and fungal pathogens) also attack Russian wheat aphid. Rolled leaves provide some protection against natural enemies. Biological control eventually may play an important role in the regulation of Russian wheat aphid populations, particularly in areas of more diversified agriculture.

Management. Cultural controls are important in management of Russian wheat aphid and other cereal pests. Generalized recommendations include good volunteer management, early planting for spring grains, and any measure that reduces crop stress. More site-specific cultural tactics include delayed planting for winter wheat, grazing, narrowed row spacing, and increased nitrogen fertilization.

Resistant wheats are being grown, particularly in parts of Colorado, but adapted cultivars are not yet available for all affected production areas. Resistant barleys and additional resistant wheats also are being developed. Resistant wheats may require insecticide applications for other pest problems. A biotype virulent to the currently deployed sources of resistance was first observed in 2003 and is now widespread in areas where resistant varieties have been grown. New sources of resistance have been identified for this and additional biotypes. In addition, there are important resistance differences among small grain species. Oats are resistant to Russian wheat aphid, and triticale is more resistant than barley or susceptible wheats.

Chemical control of Russian wheat aphid is effective, but a single insecticide application may not provide complete protection. Systemic insecticides generally are more effective than contact materials. Available systemic insecticides may be applied as seed treatments, with the seed at planting, or as foliar treatments. Control difficulties have been associated with low temperatures, poor coverage, inadequate spray volumes, failure to buffer alkaline water, excessively large aphid populations, and drought stress. Consider a contact insecticide and avoid tank mixes if the crop is drought-stressed. Foliar treatments allow the use of economic thresholds and thus the most efficient use of insect-control inputs.

The economic threshold for Russian wheat aphid in winter wheat from regrowth to heading is based on a 0.5% yield loss for every 1% infested tillers, or

$$ET = CC \times 200 / EY \times MV$$

Where:
ET = Economic threshold
 (expressed in % infested tillers)
CC = Control costs
 (cost of product and application per acre)
EY = Expected yield (bushels per acre)
MV = Market value (value per bushel)

This equation gives an infestation level, expressed in infested tillers, above which it is considered to be cost effective to make an insecticide application. Thresholds for fall infestations are more site-specific because fall infestations tend to be more frequent and economically significant in more northern regions. Thresholds for spring wheat, barley, and winter wheat infested after heading are not well defined. A sequential sampling plan (QuickSample) allows rapid assessment of the infestation level in a given field to determine whether it exceeds the calculated economic threshold.

Selected References. 106, 148, 156, 165

By Frank B. Peairs

Yellow Sugarcane Aphid

Scientific Classification. *Sipha flava* (Forbes). Hemiptera: Sternorrhyncha (Aphididae).

Origin and Distribution. Although native to North America, yellow sugarcane aphid remains endemic largely in the Gulf Coast states and Hawaii. On occasions, it may move north from the Gulf Coast states into the Missouri River valley during the warm summer months.

Description. The adult aphid is small, 1/24–1/12 in. (1.3–2.1 mm) long, oval, and hairy. Its most distinguishing feature is its color, ranging from lemon yellow to mint green. The aphids appear in two forms, apterae (wingless) and alatae (winged), depending on environmental conditions.

Pest Status. The aphid is an important pest of native and introduced grasses, sugarcane, grain sorghum, and occasionally cereal grains.

Injury. The aphid injects toxin(s) during feeding, causing a yellow-red area at the feeding site, eventually causing the infested leaves to become necrotic and eventually may kill the infested plant. Serious stand loss can occur

Pest Information

Yellow sugarcane aphids (Vladimir Beregovoy).

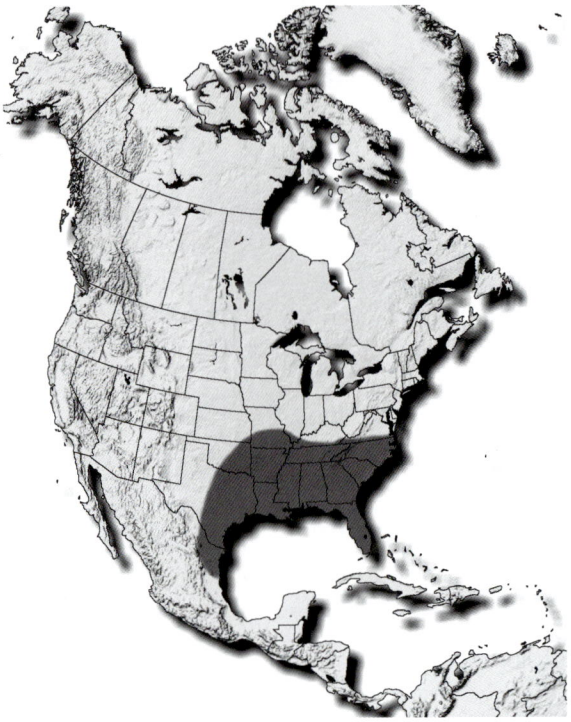

Yellow sugarcane aphid distribuition.

because damaged plants do not recover, even when plant stands are treated with a pesticide. The yellow sugarcane aphid is not known to transmit plant viruses.

Life History. Females reproduce sexually or asexually. Selected females have the ability to lay eggs (oviparity) or give birth to live young (viviparity). The combined effect of parthenogenesis and viviparity creates a rapid population buildup during times of plentiful food supply. As food reserves are depleted, the winged forms develop and disperse in search of new sources of food. There are multiple generations throughout the summer months. The yellow sugarcane aphid can survive on >50 species of Gramineae. Consequently, native and introduced grasses in the South furnish an abundance of food plants for over-wintering females.

Management. In most cases, producers do not use management tactics to control yellow sugarcane aphid on cereal grains. Insecticides approved for other cereal grain aphids may be effective against yellow sugarcane aphid; but even after treatment, plants may not recover and die. Furthermore, the common aphid parasites do not prefer yellow sugarcane aphid as a host. Barley and wheat are relatively susceptible. Oats is essentially a nonhost. Resistant germplasm has been identified in barley and wheat, but resistant cultivars are not available.

Selected References. 120, 182, 200

By S. Dean Kindler

Miscellaneous Aphids

Scientific Classification. At least a dozen other species of aphids are found on small grains, but only four are of any economic consequence. These are the western wheat aphid, *Diuraphis tritici* (Gillette); *Metopolophium dirhodum* (Walker), widely known as the rose grain aphid; the rice root aphid, *Rhopalosiphum rufiabdominalis* (Sasaki); and *Sipha elegans* del Guercio, which has no official common name. A field key that includes most of the common aphids found on cereals in North America can be found in Halbert and Voegtlin (1995). A brief discussion of each species follows.

Western Wheat Aphid. This species is native to the intermountain states and Pacific Northwest. It is shaped like Russian wheat aphid, but has no double tail (no supracaudal process). Living colonies are covered with a white wax and have an acrid odor when handled. They caused widespread damage to wheat in Montana in the early 1900s, and they occasionally cause severe localized damage today. The life cycle is similar to that of English grain aphid. A closely related introduced species, *Diuraphis frequens* (Walker), can be found more consistently than western wheat aphid in the western United States, but its

Diuraphis frequens (Susan E. Halbert).

Pest Information

severe damage is restricted to isolated plants.

Rice Root Aphid. This species is found throughout North America, but it is probably most abundant in southern states. Rice root aphids look very similar to bird cherry-oat aphids; the apterous forms have a reddish abdomen, but under magnification rice root aphids have coarse hairs (setae) on the antennae and a distinctive curve in the terminal process of the antennae. Antennae of rice root aphids also usually have 5 segments rather than the 6 segments typical of the bird cherry-oat aphid. Rice root aphids usually colonize the underground stem and roots of wheat, whereas the bird cherry-oat aphid typically forms colonies above the soil line. In the southeastern United States, rice root aphid alates are found on newly emerged wheat or oat seedlings in the fall, typically before the arrival of bird cherry-oat aphid alates. Rice root

Rose grain aphids, *Metopolophium dirhodum* (Susan E. Halbert and Guy W. Bishop).

Rice root aphids (Jay W. Chapin).

aphids probably reproduce as live-bearing asexual females year-round, but may overwinter on an unknown species of *Prunus* in the north (see bird cherry-oat aphid). Rice root aphids are highly efficient vectors of BYDV on wheat, but in studies conducted thus far, the bird cherry-oat aphid has been more important in field transmission.

Rose Grain Aphid. This European species is found primarily in western states. The aphids are straw colored or faintly green. They tend to line up parallel to the leaf veins. The life cycle is similar to that of bird cherry-oat aphids, and the overwintering host is rose. Rose grain aphids can build to large numbers on leaves of maturing wheat and barley in the Pacific Northwest. Damage is usually minimal, but rose grain aphids are implicated in BYD epidemics in Mexico.

Sipha elegans. This species is the northern equivalent of the yellow sugarcane aphid. It is an introduced Eurasian species. These aphids are most common in maturing, late-planted spring wheat, where they usually show up first as

(Top) *Sipha elegans* (Susan E. Halbert and Guy W. Bishop). **(Bottom)** *Sipha elegans* mixed with greenbugs (Guy W. Bishop).

bright yellow guests in greenbug colonies. Damage caused by *S. elegans* closely resembles greenbug damage; however, *S. elegans* usually arrive so late in the season that their damage is not economically significant. The life cycle is similar to that of the English grain aphid, but preferred hosts are desert grasses rather than small grains.

Selected References. 13, 36, 68

Armyworms

Four species of armyworms are pests of small grains in North America. Because each species has a unique life history and causes different damage, they are presented separately.

Armyworm

Scientific Classification. *Pseudaletia unipuncta* (Haworth), formerly *Cirphis unipuncta* and *Mythimna unipuncta*. Lepidoptera (Noctuidae).

Origin and Distribution. Armyworm is a native species that overwinters in the southern areas of North America and migrates north, potentially infesting grain fields in most areas of the United States and Canada.

Description. Larvae, the injurious form, develop through six stages. Mature larvae are 1 1/6–1 1/3 in. (3.0–3.5 cm) long, greenish-brown with longitudinal stripes down the back and sides. Pupae are dull brown and 5/6 in. (20 mm) long. The adult is a pale brown or brownish-gray moth with wingspan of 1 1/2 in. (3.8 cm) and a small, prominent white dot in the center of each front wing.

Armyworm (Marlin E. Rice).

Pest Status. The armyworm is a sporadic pest of feed grains throughout its range. The armyworm is a regular pest of wheat in certain areas of the mid-South. In fields that reach treatment level, the infestations are commonly 5 or more larvae per square foot. Larvae damage plants by feeding on the leaf tissue usually from the ground up. High numbers may completely defoliate fields, and larvae may feed on the stems causing the removal or cutting of grain heads. In the presence of high infestations, larvae may migrate in large numbers to nearby crops in search of food.

Life History. The armyworm overwinters as a larvae or adult in the southern areas of their range. After maturing in the spring, moths migrate northward into areas of small grain production. Pheromone traps have captured adult moths in Arkansas around March 15; light traps have been used in eastern Ontario around mid-May to capture adults. The eggs are laid in narrow bands of a few to several hundred eggs, in 2–5 rows, rarely in 1 row. Eggs may be deposited on dead and dying leaves or green leaves under the leaf sheath at the base of the plant. Eggs hatch in an average of 8 d and then larvae feed for ~3 wk. During bright daylight hours, larvae usually hide under litter on the ground. At maturity, they pupate in earthen cells 5–7 cm in the soil. Moths emerge ~4 wk later.

Management. Parasites, various diseases, insect predators, and birds attack armyworms. Blackbirds feeding in fields are a sign of armyworm in the mid-South. Many producers use this as an indicator to scout fields. Black light traps, cone pheromone traps, or both, may be used to monitor armyworm moth populations. Field scouting for infestation levels should take place from head emergence through maturity, or when armyworm infestations are normally a problem. A common method of scouting is to examine several areas of the field and express population in terms of the number of larvae per square foot. Larvae are often difficult to find because they hide under litter during daylight hours. However, on overcast days, they can often be found on the plant. Common treatment thresholds range from 3 to 6 larvae per square foot.

Selected References. 66, 85, 123

By Donald R. Johnson and Glenn Studebaker

Fall Armyworm

Scientific classification. *Spodoptera frugiperda* (J. E. Smith). Lepidoptera (Noctuidae).

Origin and Distribution. Fall armyworm is native to the Western Hemisphere and survives the winter only in semitropical and tropical zones (including the lower tip of Florida and Texas). Each year, fall armyworm moths fly north with weather fronts into the United States east of the Rocky Mountains, and west into southern New Mexico, Arizona, and California.

Description. The fall armyworm larvae have a light-colored, inverted Y on the front of a dark head. They may be black, brown, or green. Larvae have a wide, black stripe along each side. There usually are four black dots on the back of each abdominal segment. In large larvae, four distinct dots are evident on the back of the last few abdominal segments (see photo). Larvae are 1 1/4–1 1/2 in. (31.8–38.1 mm) long at maturity.

Pest Status. Fall armyworm is mainly a threat to fall-planted seedling wheat. Occasionally, fall armyworm arrives in spring wheat-growing areas while the wheat is heading and may cause economic damage.

Pest Information

Fall armyworm (Marlin E. Rice, Iowa State University).

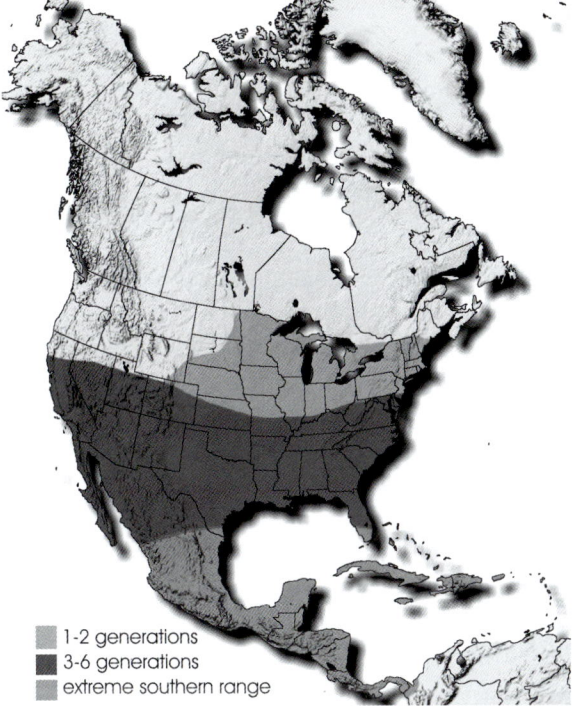

Fall Armyworm distribution.

armyworms have a wide host range, including corn, cotton, alfalfa, and pasture grasses. Populations may build on these crops and then move into fall-planted wheat. Cool, wet weather in the overwintering areas or in the Gulf States in spring is detrimental to natural enemies of the fall armyworm. This favors the development of large numbers of migratory moths. Locally, hot, dry, summers can lead to outbreaks of fall armyworm.

Management. Planting after the Hessian fly-free date often avoids damage from fall armyworms. Fields planted early for temporary winter grazing are at greatest risk of infestation. Cultivation can destroy armyworm pupae in the soil. Winter wheat fields should be scouted soon after planting for the presence of these pests. Fall armyworms are hard to control with insecticides, unless they are detected as early instars. Chemical control is probably worthwhile if there are >2 larvae per foot (30 cm) of row. If considerable damage occurs, it may be necessary to replant.

Selected References. 42, 110, 178, 197

By Kathy L. Flanders

Wheat Head Armyworm

Scientific Classification. *Faronta diffusa* (Walker). Lepidoptera (Noctuidae).

Origin and Distribution. This native pest is common in the northern United States.

Description. The larvae are tan or green caterpillars with lateral white, gray, green, or brown stripes. Adult moths have tan forewings that have a characteristic broken, darkened stripe from the basal area to the distal tips. The moths fly at night and are attracted to lights.

Injury. Larvae injure wheat by feeding on the leaves. Seedlings can be eaten down to the ground level and destroyed soon after emergence.

Life History. Fall armyworm moths are usually detected along the Gulf Coast of the United States in April, and typically move into northern areas by September. Fall armyworm has 1 generation in northern areas, and up to 6 generations in the southern United States. Fall armyworm cannot overwinter in areas with freezing temperatures. In mild winters, fall armyworm may breed continuously along the Gulf Coast.

Eggs are deposited in masses on the underside of foliage, tree limbs, and on buildings. Hairs from the female moth's body make the egg masses fuzzy. Larvae hatch in a few days and feed on a variety of host plants. Larvae burrow into the soil to a depth of 1–2 in. (2.5–5 cm) and pupate. Moths emerge to start the next generation. These

Wheat head armyworm (Wendell Morrill).

Pest Information

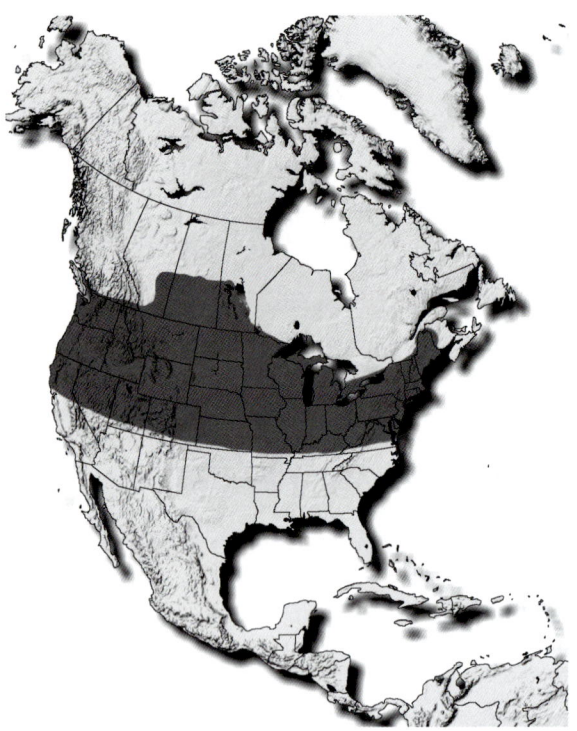

Wheat head armyworm distribution.

Wheat seed damaged by wheat head armyworm (Wendell Morrill).

Pest Status. This insect is a minor pest of wheat, although damage and infestations usually go undetected. Larvae feed on plant foliage and developing kernels.

Life History. Pupae overwinter in the soil, and moths emerge in the spring. Eggs are laid on plant foliage. Larvae feed and complete their development on above-ground plant regions. It is believed that there are 2 generations per year, although the biology has not been well documented.

Injury. Defoliation of well-established plants usually is not extensive enough to affect yields; however, larval feeding on the soft endosperm of the developing grain directly affects grain yield and quality. Damaged kernels may pass through combines during harvest. There are no good estimates of the extent of damage and, therefore, economic thresholds have not been established.

Management Practices. Light traps can be used to monitor moth activity, and larvae can be collected in sweep nets. There are no recommended management practices, but it is likely that insecticides registered for control of other armyworms or cutworms would be effective.

Selected References. 134, 176

By Wendell Morrill

Yellowstriped Armyworm

Scientific classification. *Spodoptera ornithogalli* (Guenée), formerly *Prodenia ornithogalli*: Lepidoptera (Noctuidae).

Origin and Distribution. Native to the Western Hemisphere, yellowstriped armyworm is found from Massachusetts westward to Minnesota, Nebraska, New Mexico, Arizona, and California.

Description. Yellowstriped armyworm larvae are gray to black and also have an inverted Y on the head, a yellow stripe running along each side, and a black spot just behind the hind legs. There are two dark triangular spots in the back of each abdominal segment. Larvae are 1 1/4–1 1/2 in. (31.8–38.1 mm) long at maturity.

Yellowstriped armyworm (James L. Castner, University of Florida).

Pest Status. Yellowstriped armyworm is a threat to fall-planted seedling wheat, but it is rarely of economic importance outside the southern United States.

Injury. Larvae injure wheat by feeding on the leaves. Seedlings can be eaten down to the ground level and destroyed soon after emergence.

Life History. Yellowstriped armyworm overwinters in the pupal stage in the Gulf Coast region. Moths emerge in spring and begin the first generation. There are 3–5 generations per year. Except for overwintering, life histories of the yellowstriped and fall armyworms are similar.

Management. Management tactics for yellowstriped armyworm are similar to those for fall armyworm.

Selected References. 42, 197

By Kathy L. Flanders

Billbugs

Scientific Classification. *Sphenophorus* species, *Listronotus montanus*. Coleoptera (Curculionidae, Phynchophorinae).

Origin and Distribution. Billbugs are native and present throughout grasslands and cultivated areas of United States and Canada.

Description. Adults of several species range from ¼ to 3/4 in. (6.4–19.1 mm) long, with clubbed antennae and a characteristic curved snout. Hard-bodied adults generally are reddish brown, gray, or black. Some species have longitudinal lines on the wing covers. The head is prolonged into a snout about a quarter as long as the body with chewing mouthparts at the end of the snout. Larvae occur in soil and are short, chunky, humpbacked grubs without true legs. Grubs are white with distinct, hard brown to yellowish brown head and are 1/3–1/2 in. (7.5–12.5 mm) long.

Billbug (Sue Blodgett).

Billbug injury to wheat seedling (Sue Blodgett).

Pest Status. This subfamily includes plant-feeding and grain-feeding weevils. Adults and larvae of *Sphenophorus* billbugs species use grasses and sedges as hosts and damage cultivated crops, such as small grains, rice, and corn. *L. montanus* can cause significant plant stand reduction in Montana spring wheat.

Injury. Infestation or injury by billbugs may be more common in low-lying, wet sites. Adult billbugs or snout beetles chew holes in stems of seedling plants and feed on developing plant tissue. Interior leaves or buds wilt and/or leaves are perforated with small series of holes oriented perpendicular to the leaf blade. As leaves unfold and expand, the leaf blades break at the perforations causing substantial leaf loss. Small larvae feed on leaves or within stems and larger larvae feed on grass crowns and roots. Plants may become stunted and deformed; and curling or twisting of damaged leaves can affect leaf development. Economic damage caused by this species complex is not known.

Life History. Adults overwinter in thatch, plant crowns, or cracks in the soil and disperse to grass hosts in the spring. Crop residues left on the soil surface in conservation tillage systems may favor these insects. After mating, females lay 1–2 eggs in stems or grass crowns. White, grub-like larvae feed within grass stems until constrained by size, at which time they drop to the ground and complete their feeding on grass crowns and roots. Mature larvae form earthen cells, pupating for 8–10 d in the soil. Adults emerge in late summer and fall. There is one generation per year.

Management. Pitfall traps, floating adults out of leaf litter samples submerged in water, and plant inspection have been suggested for monitoring spring adult activity. Adults actively feeding in the spring can be controlled with insecticides. Natural enemies have not been studied extensively. Parasitoids are known, but are not abundant. Crop rotation can be effective because adults are thought to travel only short distances. Crop rotation should be supplemented by cultivation that destroys volunteer host plants and disturbs crop residue where adults overwinter.

Selected References. 14, 187

By Sue L. Blodgett

Blister Beetles

Scientific Classification. Various *Epicauta* spp. Coleoptera (Meloidae).

Origin and Distribution. Blister beetles are native to North America. Although 300 species occur in North America, only ~4–5 species are found on small grains.

Description. Adults range in size from ½ to 1 in. (1.3–

Pest Information

Blister beetle (Sue Blodgett).

2.5 cm) and have a prothorax that is narrower than either the width of the head or the wing covers (elytra). Adults vary widely in color from solid gray and black to brightly spotted or striped to iridescent.

Pest Status. Although blister beetles are primarily pests of broadleaf crops, they do occur rarely on resident and cultivated grasses including oats, barley, wheat, and corn. Blister beetles are a minor pest of small grains.

Injury. Adult blister beetles cause plant damage by defoliation. Some blister beetle species tend to form dense aggregations in the field; consequently, their defoliation pattern can be patchy. Blister beetles are of particular concern because their body fluids contain cantharidin, a contact blistering agent. Livestock can be poisoned by cantharidin when they inadvertently ingest blister beetles. All blister beetles contain cantharidin in varying concentrations depending on beetle species and sex.

Life History. Eggs are laid in soil. Upon hatching, highly mobile first instars seek out a suitable host. Larvae are predatory, feeding on grasshopper eggs or on immature wild bees. Immature stages are completed within the egg pod or bee nest. Adult blister beetles emerge throughout the growing season.

Management. Blister beetles typically occur along the field edge or in one area of a field. Their large size and distinctive color make them fairly easy to detect during field visits, and sampling can be done either by inspecting plants or using a sweep net. Infestations rarely require control. Currently, no insecticides are registered in the United States for blister beetles damaging small grains.

By Sue L. Blodgett

Cereal Leaf Beetle

Scientific Classification. *Oulema melanopus* (L.). Coleoptera (Chrysomelidae).

Origin and Distribution. Cereal leaf beetle, native to Europe and Asia, was first observed in the United States in southwestern Michigan in 1962. From there, it spread to its present range, which includes nearly all states east of the Great Plains, and Idaho, Montana, Utah, and Wyoming, and Washington.

Description. The adult cereal leaf beetle is ~1/5 in. (5.1 mm) long and has a metallic, bluish-black head and wing covers. The legs and front segment of the thorax are rust red. Eggs are elliptical, ~1/32 in. (0.8 mm) long. Newly laid eggs are yellowish, but the color darkens from burnt orangish to almost black just before hatching. Eggs are most often laid singly or end-to-end in short chains on the upper leaf surface between and aligned with the mid veins. When newly hatched, larvae are very small, but they grow to a maximum size that is slightly longer than the adult. They are sluglike in shape and resemble small Colorado potato beetle grubs, except for their coloration. The head and legs are brownish black, and the body is orangish yellow. However, body coloration is usually obscured by a black globule of mucus and fecal matter held on the body, which gives larvae a shiny black, wet appearance. Pupae form into adults in earthen cells in the top 2 in. of soil. In the soil, pupae are enveloped in a thin transparent membrane. They are bright yellow at first but darken to the color of adults just before emerging.

Pest Status. Cereal leaf beetle is a widespread pest

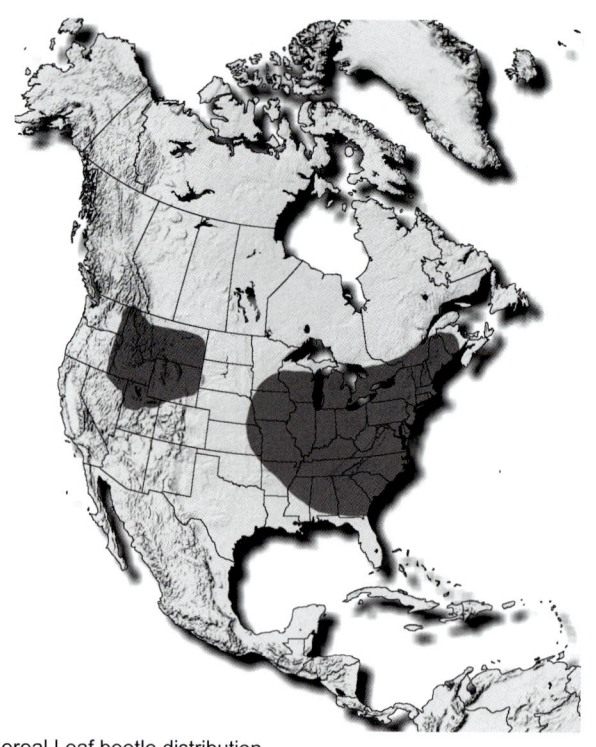

Cereal Leaf beetle distribution.

Pest Information

(Left) Cereal leaf beetle (Jerome Grant, University of Tennessee). (Center) Cereal leaf beetle larva (Wayne Bailey, University of Missouri). (Right) Cereal leaf beetle eggs (John Roberts, USDA–ARS, retired).

of small grain throughout Europe, Asia, and the United States. It feeds only on cereals and related grasses. Spring-planted oats, barley, and wheat are preferred hosts; however, damaging populations commonly occur in the early spring on fall-planted wheat, barley and rye. In the United States, throughout much of its eastern range, a complex of natural enemies introduced from Europe controls this pest. This complex consists of three larval and one egg parasitoid species. Damaging beetle populations are still encountered in some western, southeastern, and middle Atlantic states where the parasitoids have not become well established, and their controlling influence is limited or lacking. Economic damage in these areas has been attributed to several factors including unsuccessful adaptation of some parasitoid species to new climates and farmers' use of cultural practices that are unfavorable to the parasites. In the western distribution, more damage is observed in years of high rainfall. However, intense rainfall or sprinkle irrigation during egg hatch reduces larval populations and the need for control. In the East, damage is associated with late-planted, thinly sown fields. Since 1993, the USDA and several state partners have renewed efforts to locate and colonize new parasitoid species that are better adapted to conditions in states where damaging beetle populations still exist.

Injury. Damage to wheat, oats, and barley is caused by larvae feeding on leaves. The time of peak larval feeding varies somewhat with weather conditions, but generally occurs during April and May in eastern states and late May through late June in western states. Although adults will feed on young small grain plants, eating completely through the leaf and producing a striping of leaves, their feeding does not affect the plant's performance. Larvae, however, can reach very high numbers in small grains; they eat long strips of green tissue from between leaf veins and may skeletonize entire leaves, leaving only the transparent lower leaf surface tissue. These adult and larval feeding patterns are diagnostic for cereal leaf beetle. Severely defoliated fields can take on a white "frosted" cast when much green tissue is lost from upper leaves.

Often cereal leaf beetle populations will be somewhat localized, and damage will be confined to spots or sections of fields or farms. Damage can build up quickly, often in as little as 5 d, after larvae become large and warm temperatures make the insects very active. Leaf feeding reduces the plant's ability to make its food and limits reproductive growth and production of full-grain kernels, particularly if the upper leaves are destroyed. Yield reduction up to 45% has been observed when defoliation was near 100% and the damage occurred early in the heading period. Damage later in the head-fill period does not have as great an impact. Late feeding has reduced yields in Utah because it results in some partially filled kernels, which are usually blown out during threshing. Yield reductions of 10–20% have been typical in infested commercial fields in eastern states.

Life History. There is only one generation per year. Adult beetles overwinter in fallen leaves, ground litter, or other debris along hedge-rows, within wooded areas, or other protected sites in the vicinity of last season's grain fields. In the spring, adults colonize and lay eggs in small grain fields during March through mid-April in the East and April through June in the West; although it can be earlier or later and occur over longer periods depending on spring temperatures. Eggs hatch in ~5 d. Larvae develop in ~10–12 d under ideal temperatures; however, development times vary considerably during the spring. Peak larval populations usually occur in mid-April to May in the East and in June in the West. Small larvae eat a very small amount; but when full grown, larvae have voracious appetites. Upon reaching full size, the larvae dig into the ground and pupate. Pupae transform to adults in 20–25 d. A new summer generation of adult beetles emerges in late-May and June in the East and July in the West. New beetles move from small grain fields and feed on grass plants for a short period but then remain inactive through most of the summer. Because these adults need to feed before becoming inactive, they often congregate and feed in corn fields adjacent to the small grains from which they emerged. Adult feeding on corn appears like many line etchings on the blades that can concern the farmer; however, damage is usually cosmetic rather than yield reducing. Cereal leaf beetle does not lay eggs in corn. In the fall, beetles move to wooded areas, hedgerows, and ditch banks to overwinter.

Management. Successful chemical control must be directed at eggs and larvae. Treating adults has never proven effective in reducing populations. Several insecticides registered for use on small grains effectively control larvae, usually with a single, well-timed application. Recent research in Virginia and North Carolina indicates that the most effective treatments are timed to coincide with the presence of eggs and small larvae. These applications prevent larvae from reaching larger, damaging sizes. In addition, where cereal leaf beetles are present and the Hessian fly has not been a problem, avoiding late planting may be beneficial. Late-planted, thinly sown fields are more attractive to cereal leaf beetles for egg laying in the spring. Thick planted/tillered grain fields are less likely to develop high cereal leaf beetle populations. In general, following sound agronomic practices for high-yield, small grain production reduces the likelihood and impact of cereal leaf beetle. Reducing tillage also may aid in beetle management by enhancing parasite survival because deep plowing tends to bury and destroy parasite cocoons.

Field scouting should begin when egg laying is underway. Inspect individual plant stems for eggs and larvae. Samples by direct observation or sweep nets should be taken at a minimum of 10 random sites in the interior of each field (avoid the edges). Because cereal leaf beetle is often unevenly distributed in the field, it is often necessary to determine whether a portion of a field is above threshold. If the random sampling indicates an uneven distribution (many beetles in some samples but few in others), it may be necessary to subdivide the field into two or more parts and sample each part as an individual field. Economic thresholds for cereal beetle vary among states and regions. It is best to consult local Cooperative Extension or other appropriate sources for your region.

Selected References. 9, 24, 73, 76, 86

By D. Ames Herbert, Jr., John W. Van Duyn,
Michael D. Bryan and Jay B. Karren

Chinch Bug

Scientific Classification. *Blissus leucopterus leucopterus* (Say). Hemiptera: Heteroptera (Lygaeidae).

Origin and Distribution. The chinch bug is a native of North America and is distributed throughout the United States in native grassland areas. The hairy chinch bug, *B. leucopterus hirtus* Montandon, and the southern chinch bug, *B. insularis* Barber, occur in the Northeast and in the Gulf Coast states, respectively. They are mainly pests of sorghum, corn, and turf grasses but can also injure small grains.

Description. Adults are small elongate bugs ~3/16 in. (4.8 mm) long and 1/16 in. (1.6 mm) wide. They are mostly black with a white triangle in the middle of each wing making an irregular white band across the midsection of the body. Some adult chinch bugs have short wings that do not cover the rear of the body. Nymphs have no wings and are red during the first two stages. They turn black with a white band across the midbody for the last three stages, and small wing pads appear. All stages can occur in wheat simultaneously.

Pest Status. The chinch bug is rarely considered a major pest on wheat throughout its range and is usually more of a problem under dry conditions. Adult chinch bugs migrate directly into wheat in early spring. Nymphs and adults also infest other hosts after wheat or other early spring host plants begin to mature and dry up. One to many chinch bugs can infest a single plant. Injury is most likely when dry weather retards plant growth.

Injury. Chinch bugs cause damage by sucking sap from the plant base, mostly from behind leaf sheaths but occasionally from roots. Feeding punctures cause physical damage, and sap removal deprives the plant of nutrients. Symptoms include stunting with failure of leaf sheaths to properly elongate, reddish feeding marks behind leaf sheaths, and a streaking on leaves. Severely damaged plants may die or be stunted.

Life History. Chinch bugs have several generations per year: 3 or more generations in the South, 2 generations in the North. Adults spend the winter in bunch grasses, primarily of the genus *Andropogon*. In late March and early April in the South or April and May in the North, adults fly to host plants for feeding, mating, and egg laying. In the South, these adults may infest seedling corn. In other ar-

Pest Information

Chinch bug nymphs and adult (Jim Kalisch, University of Nebraska).

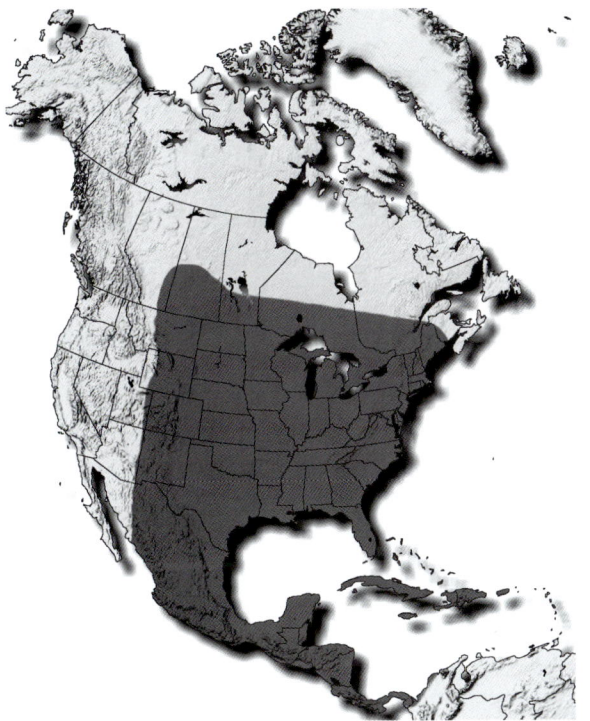

Chinch bug distribution.

eas, feeding, mating, and egg laying takes place in wheat. The growth of nymphs takes place in wheat, barley, or other small grains until the crop matures; nymphs then migrate over land up to 1/4 mile into sorghum, corn, or other grasses. Nymphs from the last generation complete their development on corn, sorghum, or wild grasses and then migrate into their bunch grasses to overwinter.

Beauveria fungi will attack overwintering chinch bugs during mild, wet winters, but its effect on overwintering populations is unknown. A small egg parasite, *Eumicrosoma benefica* Gahan, also occurs on the chinch bug, but its importance is not well defined.

Management. Although chinch bugs are common in wheat in some areas in the Great Plains, producers rarely consider it a pest. In most cases, producers do not use management tactics to control chinch bugs on small grains. Chinch bugs prefer and develop most readily in poor or thin stands. Therefore, good plant stand establishment of a small grain is important in regulating populations of this pest. Barley is preferred to wheat, but both are excellent hosts for the chinch bug. Oats is a poor host.

Data are not available for a treatment threshold for the chinch bug. When damage from chinch bugs occurs, insecticide sprays can be applied. Application by ground equipment is recommended, using high gallonage and high pressure to ensure that insecticide will reach bugs behind leaf sheaths and in the soil around the plant base.

Selected References. 123, 140, 181

By Gerald E. Wilde

Cutworms

Army Cutworm

Scientific Classification. *Euxoa auxiliaris* (Grote). Lepidoptera (Noctuidae).

Origin and Distribution. Army cutworm is native to North America and is distributed throughout the semiarid region of the Great Plains, extending to eastern Kansas and east to Illinois and Michigan. It occurs in Alberta as far north as the Peace River District and occasionally in Saskatchewan and Manitoba.

Description. Adult army cutworms range from light brown to dark grayish brown. The forewing of the adult has a prominent circular spot and kidney-shaped marking. The hind wing is grayish brown with a whitish fringe. Larvae are greenish-brown to greenish-gray, with the dorsal half being darker than the ventral half. A narrow pale mid-dorsal stripe is usually present. The head is pale brown with brown to dark brown freckles. A mature army cutworm measures 1 1/2–2 in. (3.8–5.0 cm).

Pest Status. Outbreaks of the army cutworm are sporadic and seldom, if ever, occur in any one area in large numbers on a regular basis.

Injury. Bare spots in the field in early spring may indicate cutworm activity. The army cutworm is a climbing cutworm and feeds on leaves above the soil surface. Chewed leaves are representative damage.

Life History. Army cutworm has one generation per year. Oviposition begins in late August and can extend through October. A female can deposit up to 3,000 eggs on or just beneath the soil surface. Eggs hatch in the fall following rain or snow. This cutworm overwinters in the larval stage. Larvae become active in late winter or early spring. Feeding usually occurs from dusk to dawn. This nocturnal behavior plus the small size of the early instars makes them difficult to detect even though the damage is obvious. When abundant and in short supply of food, the larvae will move en masse to adjacent fields, thus the common name. Mature larvae burrow into the soil and build earthen cells in which they pupate. Adult moths of this univoltine species emerge from these cells in May and early June. An interesting phenomenon of this cutworm species is its seasonal migration to the Rocky Mountains to avoid high summer temperatures. Adult moths return to the plains in late summer and begin egg laying.

Army cutworm is attacked by hymenopterous parasitoids belonging to the families of Ichneumonidae, Braconidae, and Chalicidae and dipteran parasitoids of Tachinidae and Bombyliidae. The following pathogens have been recorded from army cutworms: *Beauveria* spp., *Isaria* spp., *Metarhizium anisopliae* (Metschnikoff), *Sorosporella uvella* (Krassilstischik), and an entomopoxvirus (Poxviridae). Insectivorous birds are the most common vertebrate predators of larvae.

Management. Early detection of potential cutworm outbreaks is critical. Traps baited with species-specific pheromones are used during adult flights in late summer and fall for predicting the potential of a serious outbreak the following spring. Small grain fields should be monitored periodically beginning in late winter or early spring for cutworm larvae. Larval densities can be assessed by digging and screening the soil from 1 foot of row at different sites in the field. Soil samples should be dug to a depth of at least 3 in. and ~6 in. (7.6 and ~15.2 cm) on each side of the drill row. Treatment thresholds are directly related to the health and vigor of the growing crop. If the crop is experiencing moisture stress, then the effects of cutworm damage will be more dramatic. Under drought conditions, an average of 2 or more army cutworm larvae per foot (30 cm) of row may cause economic damage and justify an insecticide application. If plants are not drought stressed and appear healthy and vigorous, then the treatment threshold is raised to 4 or more army cutworm larvae per foot of row. If they are close to pupation, a treatment may not be cost-effective.

Pale Western Cutworm

Scientific Classification. *Agrotis orthogonia* Morrison: Lepidoptera (Noctuidae).

Origin and Distribution. A native species, pale western cutworm occurs from Alberta to Arizona and New Mexico, extending east to western North and South Dakota, Nebraska, Kansas, and the panhandles of Oklahoma and Texas.

Description. The adult pale western cutworm is gray to brownish white. Distinct markings on the wings are absent, although the under-surface of the wing is white. The general body color is pale yellowish-gray with a distinct white mid-dorsal line. The head is yellow-brown with two distinct vertical black dashes that form an inverted V. A mature larva measures 1 1/4–1 1/2 in. (3.2–3.8 cm) long.

Pest Status. Pale western cutworm is one of the most important pest species in the Great Plains of North America.

Injury. Bare spots in the field in early spring may indicate cutworm activity. Pale western larvae are subter-

Army cutworm (Jim Kalisch, University of Nebraska).

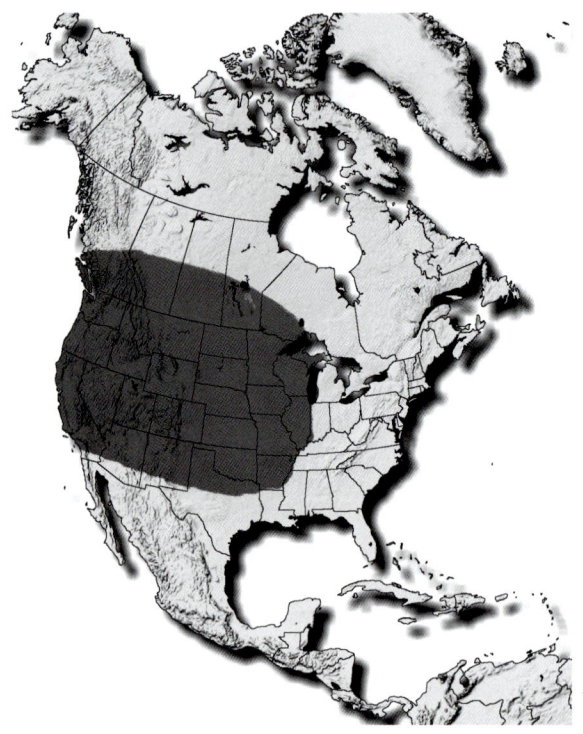
Army cutworm distribution.

Pest Information

Pale western cutworm (Jim Kalisch, University of Nebraska).

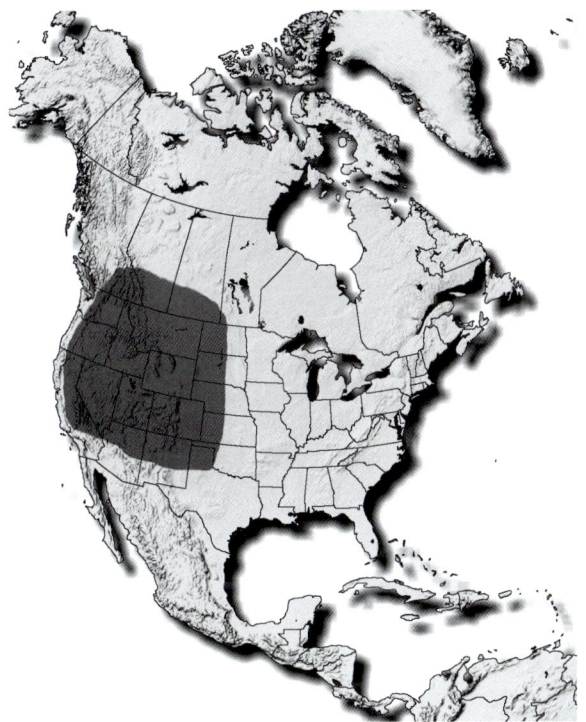
Pale western cutworm distribution.

ranean and feed on plants below the soil surface. Damage appears in wheat as dead or wilted tillers.

Life History. Pale western cutworm moths deposit up to 300 eggs per female in the upper 2 in. (5 cm) of loose soil in cultivated ground in early fall. Some hatching may occur in the fall, but most eggs hatch the following spring. The newly hatched larvae are small and difficult to detect. Each instar is subterranean and feeds on plants below the surface of the ground. Mature larvae become less active and burrow vertically into the soil, where they form an earthen cell. Here, they pass most of the summer in a prepupal state and then pupate in August. Adult moths emerge from the cell during late summer, mate, and lay eggs soon after. The prepupal and pupal periods are longer in the southern part of the range, so that a single generation occupies a full year throughout its distribution. Whereas dry periods during August through October are detrimental to egg hatch and larval survival of army cutworms, dry weather favors an increase in pale western cutworms. Excessive precipitation in the spring can drown larvae in low-lying areas, force larvae to the surface of the soil where they are exposed to attack by predators and parasites, and produce conditions favorable for the increase of pathogens. A method of predicting outbreaks is based on the number of "wet days" (i.e., days on which 0.25 in. or more precipitation falls) in May and June. More than 10 wet days result in a decrease in numbers of cutworms, whereas <10 d is followed by an increase.

Pale western cutworm is attacked by hymenopterous parasitoids belonging to the families of Ichneumonidae, Braconidae, and Chalicidae and dipteran parasites of the families Tachinidae and Bombyliidae.

Management. Early detection of potential cutworm outbreaks is critical. Traps baited with species-specific pheromones are used during adult flights in late summer and fall for predicting the potential of a serious outbreak the following spring. Small grain fields should be monitored periodically beginning in late winter or early spring for cutworm larvae. Larval densities can be assessed by digging and screening the soil from 1 foot of row at different sites in the field. Soil samples should be dug to a depth of at least 3 in. and ~6 in. (7.6 and ~15.2 cm) on each side of the drill row. Treatment thresholds are directly related to the health and vigor of the growing crop. If the crop is experiencing moisture stress then the effects of cutworm damage will be more dramatic. Depending on the health of the wheat, the treatment threshold for pale western cutworms is 1–2 larvae per foot (30.5 cm) of row. Before making a decision to treat a field, it is important to determine the size of the larvae and determine whether they will pupate soon. If they are close to pupation, a treatment may not be cost-effective.

Selected References. 26, 88

By Gregory Johnson

Flea Beetles

Scientific Classification. Various species, Coleoptera (Chrysomelidae).

Origin and Distribution. Flea beetles occur throughout most small grain production areas in North America.

Description. Adults are ~1/8 in. (~3.2 mm) in length, solid black or dark gray or sometimes black with thin, pale, longitudinal stripes. Eggs are minute and yellow-white. Larvae are white and grublike with a distinct head

Pest Information

but indistinct legs.

Pest Status. Flea beetles are minor pests of small grains.

Injury. The damaging stage is the adult flea beetle. Adults chew small circular holes in leaves or elongated holes that skeletonize leaves. Damage is usually confined to the edge of fields and is worse under cool damp conditions where plant growth is slow. Flea beetles usually are a concern when feeding on seedling plants.

Flea beetle (Keith S. Pike).

Life History. Flea beetles that defoliate small grains typically have a broad host range and opportunistically feed on cereal crops. Eggs are laid in soil, and larvae feed on plant root hairs and root tips. Larval root feeding is not thought to be important in small grains. Adults emerge from soil and feed on host plant foliage. Generally, adults overwinter at the base of plants in field edges or wooded areas. Adults move into fields in the spring and may be concentrated along the field edge.

Management. Small grains are very tolerant of defoliation, hence flea beetles rarely require active management. Some states recommend controlling flea beetles when damage is extensive or when damage to seedling exceeds 50% defoliation, and beetles are present. If beetles are concentrated along the field margin, it may only be necessary to treat the field edge.

Selected References. 72

By G. David Buntin

Frit Fly

Scientific Classification. *Oscinella frit* (L.). Diptera (Chloropidae).

Origin and Distribution. Frit fly occurs throughout North America, Europe, and parts of Asia.

Description. The shiny black adults are ~1/8 in. (~3.2 mm) long. Eggs are minute, elongate, and white. Larvae are maggots with no distinct head or legs. Newly hatched larvae are white, but become yellowish when full grown.

Frit fly adult and larva (Mike P. Tolley, Dow AgroSciences).

Pest Status. Frit fly is an important pest of oats and other small grains over most of Europe. Frit fly generally is an insignificant pest in North America, but it is found more regularly on wheat in the north-central and eastern United States.

Injury. During vegetative growth stages, larvae mine leaves and later bore into stems, where they feed on the growing point of the shoot. This can distort or kill the infested tiller. During jointing, larvae feed on the developing grain spike, which can cause sterility. Spring small grains generally are more seriously damaged than winter small grains.

Life History. Biology of the frit fly has not been extensively studied in North America. Frit fly usually has 3 generations per year. It attacks all small grains and several other cultivated and wild grasses. Larvae overwinter, and pupation occurs in the spring. In winter-type small grains, the first and second generations feed in the developing stem and grain spike of winter crops, with larvae of third generation mining leaves and feeding on the growing point of tillers in the fall. In spring-type small grains, larvae of the first and second (if it occurs) generation mine leaves and feed on the growing point of tillers of vegetative stage plants, whereas the later generation attacks jointed stems.

Management. Frit fly is seldom actively managed in North America. In spring-type small grains, early planting may reduce infestation. Seed treatment and foliar-applied insecticides are sometimes used to manage this pest in Europe.

Selected References. 3, 72

By G. David Buntin

Grasshoppers

Scientific Classification. Redlegged grasshopper, *Melanoplus femurrubrum* (De Geer); twostriped grasshopper, *M. bivittatus* (Say); differential grasshopper, *M. differentialis*

(Thomas); migratory grasshopper, *M. sanguinipes* (F.); and clearwing grasshopper, *Camnula pellucida* (Scudder): Orthoptera (Acrididae).

Origin and Distribution. All of the pest species of grasshoppers in North America are native. They are widely distributed throughout North America; however, seldom is damage widespread east of the Mississippi River. Although there are >600 species of grasshoppers in North America, only a few are serious economic pests. The redlegged and migratory grasshoppers are the most widespread species in North America, followed by the twostriped, differential, and clearwing grasshoppers. At least 1 or 2 species occur in every wheat-growing region.

Description. All species of grasshoppers have enlarged hind legs and the major pests of small grains have functional wings. Females are slightly larger than the males in all species. The *Melanoplus* spp. are referred to as the spur-throated grasshoppers because they have a spur between their front pair of legs. The clearwing grasshopper is a banded wing species and has clear forewings with dark mottled patches.

The redlegged grasshopper is 3/4–1 in. (1.9–2.5 cm) long and bright yellow on the underside with bright red hind tibia. The migratory grasshopper is 7/8–1 1/8 in. (2.3–2.8 cm) long with a tan to gray body; the hind tibia is blue green or red. The differential grasshopper is 1 1/2–1 3/4 in. (3.8–4.5 cm) long and has a yellow-green body and chevron-like stripes on the outside hind femurs. The twostriped grasshopper is 1 1/2–1 7/8 (3.8–4.8 cm) long, has a green-tan body and two, light yellow dorsal stripes. The adults of the clearwing grasshopper are 3/4–1 1/2 in. (1.9–2.5 cm) long and yellow to brown.

Pest Status. Grasshopper may infest many crops, including corn, alfalfa, clover, cotton, small grains, sugarbeet, soybean, sunflower, potato, and rangeland. Areas that are extremely susceptible to grasshopper outbreaks receive <30 in. (75 cm) of precipitation in a year, with outbreaks occurring during drought cycles. Damage caused by nymphs tends to be more severe on the field edges because young flightless grasshoppers migrate in from adjacent land. Damage caused by adults is not necessarily limited to field edges because the adults are winged and can migrate. Because drought conditions can be damaging to crop yield and promote large populations of grasshoppers, crop losses are usually more severe during drought.

Injury. Adults and nymphs can injure small grains by feeding on leaf, stem, or reproductive tissue. During outbreaks, entire fields may be denuded. Grasshoppers feed by tearing off plant tissue with their mandibles and consuming the parts they remove, giving the plant a ragged

Redlegged grasshopper (Marlin E. Rice, Iowa State University).

Differential grasshopper (Marlin E. Rice, Iowa State University).

appearance. During their feeding process, grasshoppers waste more tissue than they consume.

Life History. The life histories of all the pest grasshopper species are similar. The species that injure small grains have a single generation per year and overwinter in the egg stage. The female oviposits in the soil by forming a burrow with her abdomen. She deposits a cluster of eggs 1/2–2 in. (1.2–5.1 cm) below the soil surface, and then secretes a frothy covering over the eggs. Soil particles adhere to the frothy secretion. The secretion hardens, forming a soil-covered egg pod ~1 in. (2.5 cm) long. Egg pods contain from <10 to >100 eggs. In general, a single female can deposit an average ~200–300 eggs in her lifetime. Most species deposit eggs in uncultivated fields, roadsides, and pastures. These areas are called "egg beds", and egg densities in them can be extremely high during outbreak years.

The eggs develop in the fall until cold temperatures stop development. Eggs that are deposited early in the fall will be more advanced in development than those laid later; thus, the eggs deposited early in the fall will hatch earlier in the spring. Egg hatch can begin in late April, but may be as late as early July. In general, a warm spring following a warm fall favors an early hatching date, whereas a cool fall and spring delays the hatching date.

Nymphal development of all species consists of 5–6 instars. The first instar is very small, about the size of a kernel of wheat, 1/8 in. (3.2 mm). With adequate food and warm, dry weather conditions, nymphal development takes 35–50 d.

Management. Weather can directly regulate grasshopper populations, as well as the effect grasshoppers may have on the developing crop. The previous summer's temperatures can have a pronounced influence on the current season's grasshopper population. A large number of warm, sunny days from late June through September allow adults to develop quickly and allow females to deposit large numbers of eggs. Conversely, numerous cool days through the same time period result in fewer eggs. Extreme, prolonged drought can cause eggs to desiccate or stop egg development completely.

Natural enemies can also be extremely important in regulating grasshopper populations. Bee flies (Bombyliidae), blister beetle larvae (Meloidae), ground beetles (Carabidae), and crickets are predaceous on grasshopper eggs. Predation rates of >50% have been observed in localized areas. Spiders, birds, and robber flies (Asilidae) prey upon nymphs and adults. Several pathogens and parasitoids attack nymph and adult grasshoppers. Epidemics of the fungus *Entomophaga grylli* occur when environmental conditions are warm and humid. A microsporidian *Nosema locustae*, which occurs naturally and is commercially available, can also reduce grasshopper populations.

Soil tillage has been recommend for several years as a means of destroying eggs and exposing them to various predators. However, control of grasshoppers by tillage may not be related directly to egg destruction. Tillage also destroys weeds and volunteer crops that the first instar nymphs require for survival. First instars rarely move >1 yd (0.91 m^2) from their hatch site, and if living plants are not available, the nymphs starve. Therefore, for tillage to be effective, all living plants must be destroyed before egg hatch. Third instars can walk long distances, so tillage is no longer effective. Moldboard plowing or any tillage operation that buries eggs to a depth that prevents hatching nymphs from emerging from the soil reduces populations. However, soil erosion and reduced soil moisture must be weighed against the benefits of tillage.

Specific economic thresholds for grasshoppers attacking small grains vary. Treatment guidelines vary, and specific local recommendations can be obtained from county or regional extension offices, crop consultants, or state extension entomology specialists. Generally, control using insecticides is more effective if started early in relationship to grasshopper development. With an early application, less insecticide is required to provide control; and usually, a smaller area needs to be treated because grasshopper populations develop outside of the field (e.g., road ditches). Several insecticides are labeled for control of grasshoppers in small grains.

Selected References. 30, 124, 152

Michael J. Weiss

Hessian Fly

Scientific Classification. *Mayetiola destructor* (Say). Diptera (Cecidomyiidae).

Origin and Distribution. The Hessian fly has existed in southern Europe for many centuries and probably followed wheat from its original habitat in southwest Asia. This major wheat pest is now widely distributed throughout most wheat-growing regions of Europe, North Africa, Asia, New Zealand, and North America. There is no record of its occurrence in southern or southeastern Asia, Mexico, or South America. In the United States, it occurs in all major wheat-growing areas from the Atlantic Coast to the Great Plains. In the western United States, it is found in parts of California, Idaho, Montana, North and South Dakota, Oregon, and Washington. In the semiarid regions of the Northwest, the Hessian fly is more widely distributed and abundant on irrigated wheat.

Description. The adult Hessian fly is small, dark gray to black, and ~1/8 in. (3.2 mm) long. The egg is glossy red, cylindrical and ~1/50 in. (0.5 mm) long. The first instar is also red, but turns white within a few days. The cuticle of the first instar stretches considerably during development, making the older first instar appear more robust and sluglike than the newly hatched larva. The second instar is maggotlike in appearance, white and ~1/6 in. (4.2 mm) long when mature. The body of the second instar is unevenly cylindrical, reflecting the shape of its restricted feeding site on the plant. Crowded specimens of second instars can be considerably and unevenly compressed. The puparium, commonly called "flaxseed" because of its resemblance to the seed of flax, is dark brown and 1/8–1/5 in. (3.2–5.1 mm) long.

Pest Status. The Hessian fly has a long history of pestilence in North America, but its pest status is not known in most other regions of the world, with the exception of North Africa, where it has been reported as a serious pest of wheat in Morocco, Algeria, and Tunisia since the early 1900s. In the United States, widespread outbreaks have

Pest Information

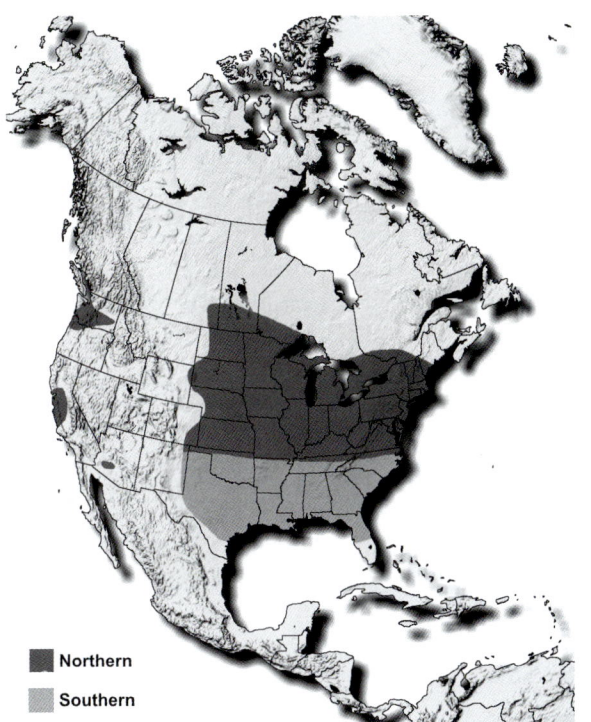

Hessian fly distribution.

Hessian fly adult (Jerome Grant, University of Tennessee).

Hessian fly eggs on wheat leaf (Roger H. Ratcliffe).

Hessian fly larvae and puparia (Roger H. Ratcliffe).

occurred at irregular intervals; however, local outbreaks can cause extensive crop losses almost every year. Wheat is the preferred host of the Hessian fly, although barley, rye and triticale are other cultivated cereal crops that occasionally are infested. Several wild grasses may serve as hosts when these crops are not available.

Injury. The Hessian fly is capable of infesting and injuring all classes of wheat grown in the United States. Feeding injury is caused entirely by the larvae. On young plants, the larvae feed beneath the leaf sheath at the base of the plant. The feeding mechanism of the larva and how it obtains food from the plant are not fully understood. A study of the larval mouthparts has revealed highly specialized mandibles that are probably used to inject salivary fluids into plants. These secretions are believed to contain enzymatic substances that inhibit plant growth and increase cell wall permeability, allowing the larva to suck the juices from the plant. No injury to plant tissue has been observed at feeding sites, although infested plants show a characteristic stunted appearance. The number of stems and leaves and weight of leaves and roots are reduced in wheat plants infested with Hessian fly larvae. Leaves of infested plants also appear more erect and are shorter and darker green than those of uninfested plants. When infestations are severe, stunting of seedlings occurs earlier, and many of the young plants die after the larvae have matured. Larval feeding also reduces the winter hardiness of wheat plants that survive the fall infestation and contributes to further loss during the winter. Heavily infested fields show areas of dead plants and thin stands. At lower infestation levels, seedlings may survive and develop new tillers that allow the plant to grow despite the loss of the main stem.

In the spring when plants are in the jointing stage, the larvae feed just above the node between the leaf sheath and stem. Feeding causes injury that prevents normal elongation of internodes and transport of nutrients to the developing spike. This injury reduces quantity and quality of the grain. The most obvious damage occurs when infested culms break at the weakened nodal areas; much of the grain with broken culms is lost when the crop is harvested.

Wheat seedlings injured by Hessian fly; note dark green leaf color of injured plants (David Buntin).

Life History. In winter wheat areas, the typical life cycle of the Hessian fly begins with fall emergence of adults from infested wheat stubble or volunteer wheat. Soon after mating, the female begins laying 200–300 eggs on the upper leaf surface of young plants. The adults do not feed, are short lived, and die within a few days after emergence. The Hessian fly female tends to lay more eggs at higher humidity; survival of small larvae is higher under these conditions, and causes higher levels of infestation. Eggs hatch in 3–10 d depending upon temperature. Newly hatched larvae crawl behind the leaf sheaths and migrate to the crown of the plant where they begin to feed. Only the first instar is mobile, but first and second instars feed and grow.

The larval developmental time varies, but most larvae are full grown before the onset of cold weather. When the second instar completes growth, the outer skin hardens and forms a protective puparium in which the third instar and pupa subsequently develop. The insect overwinters in the puparium. The following spring the pupa completes development, and adults emerge and infest wheat about the time the plants begin to joint. Most of the larvae of the spring generation are found just above the nodes under the leaf sheath. The larvae pass the summer inside the puparia in the dry stubble. In late summer or fall, the larvae pupate, and adults emerge and infest volunteer or early seeded wheat. In northern winter wheat areas, two generations usually are produced each year, whereas in southern areas, supplementary broods may develop either before or after the main fall or spring generation. In the northern spring wheat areas, only one annual spring generation is produced.

Hessian fly infestation in seedlings can be diagnosed by pulling up plants and examining the dead or stunted stems. Leaves should be peeled down to their points of attachment with the stem. The larva or flaxseed is located at the feeding site above the point of attachment. In older plants, the immature insect can be found within the shortened internode at the point of infestation or in lodged stems, just above the node where the stem broke.

Management. Many biotic and abiotic factors regulate the abundance and destructiveness of the Hessian fly. When these factors favor the insect's survival and development, populations may increase rapidly from one generation to the next. Consequently, when or where economic infestations may appear is not always predictable. For this reason, most control methods for the Hessian fly are preventive rather than remedial. The three most important methods for reducing infestation are planting resistant wheat varieties, delayed seeding of winter wheat to escape fall infestation, and destroying volunteer wheat. Delayed fall planting, or planting after the "fly free" date, is used to escape the peak emergence of adults and is adjusted according to the mean temperature for the wheat-growing area. Dates range from mid-September in the upper Midwest and northeast, to early October in the Northwest, and to late October in the Southeast. When remedial action must be taken to suppress heavy Hessian fly infestations, systemic insecticides applied at planting provide effective protection of wheat from fall and winter generations.

Several parasitoids attack the Hessian fly, but there has been no good assessment of their effect on suppressing fly populations. The level of parasitism is often very high at the end of the season. Parasitoids contribute to the reduction of the population for the next season, but offer no protection for the current crop, which has been damaged by the time they become an important factor. No concerted effort has been made recently to apply biological control measures in management of the Hessian fly.

In the following discussion, Hessian fly management is divided rather broadly between northern and southern wheat-growing areas. The life history of the insect and, in some cases, the class of wheat grown within these areas may influence the effectiveness of specific management practices.

Northern United States. This area includes approximately the northern two-thirds of the United States and Canada including the Great Plains spring wheat regions and the western winter and spring wheat region, and the northern half of the winter wheat region in the Midwest and eastern United States.

Resistant wheat varieties provide the most reliable and economical control of the Hessian fly in much of the northern soft winter wheat region of the eastern United States and the Great Plains winter wheat-growing region. There are fewer Hessian fly-resistant spring wheat varieties. Resistant sources and their genes incorporated into new wheat varieties. However, growing wheat varieties with high levels of resistance exerts a strong selection pressure on Hessian fly populations that favors races, or biotypes, capable of surviving and reproducing on resistant wheat. Therefore, the biotype composition of Hessian fly populations within wheat-growing regions changes with exposure to wheat varieties carrying specific resistance genes; over time, the effectiveness of varieties in preventing infestation by specific biotypes is reduced. Varieties with genes from new sources of resistance are continually developed and released when older varieties lose effectiveness. Sixteen Hessian fly biotypes, classified as 'GP' (Great Plains) and 'A' through 'O', have been identified on the basis of their differential response to the resistance genes H3, H5, H6, and H7H8 in wheat. The Great Plains biotype is considered to have been predominant in Hessian fly field populations before exposure to resistant wheat varieties and subsequent selection for virulence to resistance genes in these varieties. Hessian fly biotypes 'A' through 'O' differ in the number of resistance genes to which they express virulence. Biotype 'L' is the most virulent and can attack wheat varieties with any of these genes. Currently, Biotype 'L' is common in Hessian fly populations in much of the U.S. soft winter wheat region of the Midwest and East.

The greatest change in Hessian fly biotype composition has occurred in the soft winter wheat region of the eastern United States, with major shifts in biotype composition and virulence to resistance genes in wheat occurring after resistant varieties are widely grown for 6–8 yr. Much less change in biotype composition has occurred in Hessian fly populations exposed to hard red winter wheat in the Great Plains.

Resistant (left) and susceptible wheat after Hessian fly damage (David Buntin).

Delayed fall planting of winter wheat and destruction of volunteer wheat are recommended management practices to reduce Hessian fly infestation in the northern United States. In areas where winter and spring wheat are grown and Hessian fly infestation is more severe, winter wheat generally incurs less damage and is recommended for planting wherever a choice is possible. Destruction of volunteer wheat is a recommended cultural practice in winter and spring wheat production areas. Insecticides are seldom used in much of the northern wheat-growing area for Hessian fly control.

Southern United States. In the southern United States, delayed planting of winter wheat is a less effective for managing the Hessian fly because adults oviposit and larvae development can occur through the winter. In this area, destruction of volunteer wheat, which serves as a breeding place for Hessian fly in the late summer, and tillage to bury stubble are the most useful cultural control approach.

Systemic insecticides, applied at planting, are effective in protecting winter wheat from fall and winter generations of the Hessian fly in the southern United States. An at-planting treatment also suppresses the first spring brood indirectly by controlling the previous 1–2 broods within that field, but it does not provide enough residual activity to control flies moving from adjacent areas in the spring. Foliar treatments of systemic insecticides timed to control the spring broods are not effective and are not recommended. Granular in-furrow treatments of systemic insecticides are more effective than broadcast or banded liquid application treatments applied with a boom mounted in front of the wheat drill. Newer systemic seed treatments also control fall infestations.

Hessian fly–resistant wheat varieties should be grown when available. As described for the northern soft winter wheat region of the eastern United States, Hessian fly biotype composition has changed significantly in much of this area since the 1980s, as Hessian fly populations have responded to exposure to resistance genes in improved wheat varieties. State variety recommendations should be followed to identify the best sources of resistance. When available, resistant varieties can be particularly useful in avoiding the use of insecticides on early planted, high-risk fields.

Selected References. 22, 71, 157, 169

By Roger H. Ratcliffe

Leaf Sawflies

Scientific Classification. The most important species are *Pachynematus sporax* Ross, *P. setator* Ross, *Dolerus sericeus* Say, and *D. nitens* Zaddach. Hymenoptera (Tenthredinidae). Additional species in both genera either have been recorded from wheat or are potential feeders on grain crops.

Origin and Distribution. All species are native to North America, except *D. nitens*, which is introduced from Europe. Both genera are transcontinental in the United States and southern Canada to Alaska. Certain species are more restrictive, e.g., *P. sporax* occurs from British Columbia to California, and *D. sericeus* occurs in eastern North America. At least one species occurs in most wheat-growing areas.

Description. Adults are 1/4–1/3 in. (6.4–8.5 mm) long. *Dolerus* spp. are black or black and red. Females of *Pachynematus* spp. are mostly yellowish brown with various dorsal black markings on the head and body; males are mostly black. Larvae of both genera are green with various black spots or longitudinal stripes. Larvae of *Dolerus* have prolegs on abdominal segments 2–8 and 10, and six annulets on abdominal segments 1–8. Larvae of *Pachynematus* have prolegs on segments 2–7 and 10, and 5 annulets on abdominal segments 1–8.

Pest Status. Population levels are usually very low. Infestations are sporadic and of short duration, and there are only occasional reports of extensive damage. The most serious outbreak was by *P. sporax* in California where, in the early 1950s, ~200 acres of wheat were infested. Occasional damage by *P. setator* and *D. nitens* has been reported on fescue grasses grown for seed in Oregon.

Injury. Damage is caused mainly by larvae feeding on the foliage. They also may feed on the stalk and cut off the heads of wheat. Larvae of *P. sporax* can reduce plants to 2 in. (5.1 cm) stubble. *Dolerus* larvae have been observed cutting off wheat ~4 in. (10.2 cm) below the head.

Leaf sawfly larva; note the large number of prolegs that distinguish sawfly larvae from moth larvae (Ken Gray, Oregon State University).

Leaf sawflies on wheat (Mary K. Corp, Oregon State University).

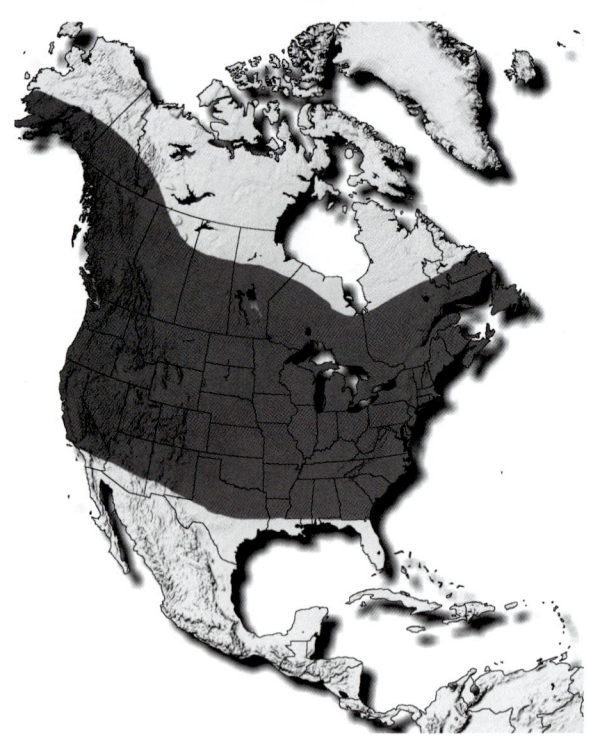
Leaf sawfly distribution.

Life History. Species of both genera have one generation a year. Adults appear early in spring, from March through May. Eggs are inserted in the leaves. Larvae feed in the spring. At maturity, larvae form a cell in the soil and remain as prepupae through the summer and winter. Pupation and adult emergence occurs the following spring.

Management. Leaf sawflies are sporadic and very infrequently encountered in small grains. When present, they rarely reach pest status or cause damage. Consequently, management rarely is needed, and insecticidal control is not recommended.

Selected References. 7, 91, 171

By David R. Smith

Leafhoppers and Planthoppers

Scientific Classification. Various species including the painted leafhopper, *Endria inimica* (Say); aster leafhopper, *Macrosteles quadrilineatus* Forbes; gray lawn leafhopper, *Exitianus exitiosus* (Uhler), and others. Hemiptera: Auchenorrhyncha (Cicadellidae). Planthoppers, *Sogata* spp. Hemiptera: Auchenorrhyncha (Delphacidae).

Origin and Distribution. Leafhoppers and planthoppers that infest small grains are native pests mainly of grasses. The painted leafhopper and gray lawn leafhopper occur throughout North America. The aster leafhopper overwinters in southern areas and migrates north into the northern United States and Canada each season.

Description. Leafhopper adults are 1/16–1/8 in. (1.6–3.2 mm) long, have a rounded or sometimes pointed head, and are often tapered distally. The painted leafhopper is grayish yellow with dark brown markings and a pair of black spots on the face, pronotum, and scutellum. The aster leafhopper is greenish yellow with 3 pairs of spots or dashes in the vertex (face). Nymphs resemble the adults with late instars having wing pads.

Pest Status. Leafhoppers and planthoppers are minor pests and usually not economically important in most areas. American wheat striate mosaic virus, vectored by the painted leafhopper, is important some years in isolated areas of the northern Great Plains along the United States–Canadian border.

Injury. Leafhoppers have piercing–sucking mouthparts and extract plant sap. Direct-feeding injury usually is not important, although large numbers can wilt and discolor foliage. Leafhoppers mainly cause injury by vectoring plant viruses. The painted leafhopper is the primary vector of American wheat striate mosaic virus in wheat in the Dakotas and Canadian prairie provinces. Aster leafhopper vectors oat blue dwarf virus to oats. The gray lawn leafhopper is not reported to vector plant disease agents. *Sogata* spp. planthoppers also transmit rice hoja blanca virus to wheat in rice-growing areas of the southern United States.

Life History. Leafhoppers lay eggs in clusters inserted

Leafhopper adult (Keith S. Pike).

in wheat leaf tissue. Five nymphal instars feed on the host plant. Each species has 1–2 generations in the North and several generations in the South. Leafhoppers that infest small grains have broad host ranges and attack numerous cultivated and wild grasses. The painted and gray lawn leafhoppers are quite common on pasture and turf grasses. Leafhoppers acquire virus by feeding on infected plants and can remain infectious for many days. The viruses are not passed to the eggs. Spiders, predatory bugs, and lady beetles consume leafhoppers and planthopper nymphs; and dryinid wasp and strepsipteran parasitoids attack nymphs and adults. However, these natural enemies usually do not regulate infestations.

Management. Leafhoppers and planthoppers usually are present in low numbers in most small grain fields. Typically, they are not economically important pests and are not actively managed. Wheat cultivars with moderate tolerance to American wheat striate mosaic virus have been reported. Wheat sown in late autumn and early spring may avoid large leafhopper populations and thereby escape infection. Insecticides may control large infestations that cause direct damage, but viral disease suppression by insecticidal control of leafhoppers probably is not cost-effective.

Selected References. 45, 145, 204

By G. David Buntin

Leafminers and Grass Sheathminer

Scientific Classification. Grass sheathminer, *Cerodontha dorsalis* (Loew), formerly *Odontocera dorsalis* Loew and *C. femoralis* (Meigen); and *C. occidentalis* Sehgal. Diptera

Pest Information

(Agromyzidae).

Origin and Distribution. The grass sheathminer is widespread in North America and is likely present in all U.S. states and most of Canada. It has been reported in Mongolia, but it also occurs in Argentina, Brazil, Peru, Guatemala, Costa Rica, Dominica, and Puerto Rico. *C. occidentalis* is a potential pest that occurs from Alaska to southern Canada and south into most western U.S. states.

Description. Adults of both species are yellow and black with considerable color variation. The third antennal segment has a prominent spine beneath the bristlelike arista. The grass sheathminer is small 1/16–1/10 in. (1.6–2.5 mm) long with a wing length of ~1/10 in. (2.5 mm). The eastern and southern form is pale overall. The western form is blackish with yellow restricted to the head, sides of thorax and on the legs; the thorax above is grayish-black matte with a prominent yellowish-gray patch. The two forms overlap in Colorado. Adults of *C. occidentalis* are similar but slightly larger than *C. dorsalis*, with a dark or black thorax. Eggs are white and oval. Larvae are white to yellowish maggots, 1/6–1/4 in. (4.2–6.4 mm) in length. Pupae (=puparium) are shiny yellowish brown and somewhat flattened.

Pest Status. Leaf and sheathminers, especially *C. dorsalis*, may be of some economic importance, but their significance needs to be established. Despite initial damage to young plants, they seldom suffer permanent or severe injury. Healthy, vigorous plants often outgrow leaf-mining damage. It is uncertain whether there are any resulting reductions in yield. Host plants are a wide range of grasses including many wild and most cultivated genera and cereal crops.

Injury. Adult females make many feeding and oviposition puncture wounds in leaf surfaces resulting in loss of sap and progressive yellowing of leaves from tip to base. Larval mining and feeding within leaves create the most damage. Severed or heavily mined leaves may wither and drop away. In young plants, larvae occasionally mine from the leaf sheath into the plant stem near or below the ground and may kill the plant.

Life History. Information is available for *C. dorsalis*, but it probably applies to other grass leafminers as well. Adult flies insert eggs beneath the leaf epidermis, usually on the upper sides toward the tips. Larvae emerge and feed rapidly as they tunnel down the leaf toward the leaf sheath. The length of larval lifespan is 9–24 d, but it varies with season and locality. The pupal stage occurs internally within the leaf sheath and lasts 9–24 d. In California, up to 8 generations are reported with complete overlap so that all stages can be found simultaneously throughout the winter.

Wheat leaf with mine (Keith S. Pike).

Management. In most cases, management tactics are not used to control grass leaf and sheathminers on small grains. Insecticides are not recommended. Natural enemies, including nine parasitoid wasps in the United States, probably control populations in most situations.

Selected References. 2, 112, 179, 180

By William Turner

Lesser Cornstalk Borer

Scientific Classification. *Elasmopalpus lignosellus* (Zeller); formerly *Pempelia lignosella, Dasypyga carbonella, E. angustellus,* and *E, puer.* Lepidoptera (Pyralidae).

Origin and Distribution. A native species, widely distributed across North and South America that has been accidentally introduced in many other parts of the world.

Description. Larval stage is injurious and has 6 instars. Mature larvae are bluish with dark transverse bands and are >5/8 in. (15.9 mm) long. Larvae inhabit the upper soil and construct silken tubes covered with soil particles that are attached to the wheat or other small grain stem. Pupae are initially greenish and later become brown. Adults are 1/2 in. (12.7 mm) long, and the wings are folded together at rest. Adult males are tan with a dark stripe down the center of the back, and adult females are uniformly dark. Wing color and linear shape are diagnostic characters.

Pest Status. The lesser cornstalk borer is common in

the early fall in the southern United States and can sometimes be a pest of wheat, oats, and rye planted in October for winter grazing. Problems occur only during warm, dry conditions.

Injury. Larvae bore into the stem at or just below the soil surface. A single larva can injure many seedlings. Seedlings may be cut off at the soil surface or may wilt and die.

Life History. Lesser cornstalk borer females lay eggs singly or in small groups on the soil surface or directly on the stem near the soil surface. Eggs are greenish white and turn pink to red before hatching. Larvae feed on seedlings, older plants, or decaying plant residues. The lower developmental threshold is 58 °F (14.4 °C) and egg-to-adult development requires 791 Fahrenheit degree days (FDD) or 439 centigrade degree days (CDD). The preovipostional period requires 27 FDD (=15 CDD). Adults live 275 FDD (=153 CDD), which enhances the overlap of generations. The number of generations per year varies with latitude. Five or more broadly overlapping generations occur in southern latitudes, and all life stage may overwinter. Numerous species of parasitoids, pathogens, and predators kill lesser cornstalk borers. Parasitoids include the parasitic wasps *Orgilus elasmopalpi* (Muesebeck), *Pristomerus spinator* (F.), and *Chelonus elasmopalpi* McComb, and the tachinid fly *Stomatomyia floridensis* Townsend. Predators include the striped earwig [*Labidura riparia* (Pallas)] and red imported fire ant (*Solenopsis invicta* Buren). Pathogens include a granulosis virus and a species of *Beauveria* fungus.

Management. Only fields planted in October are at risk, and the damage potential is greatest under warm, dry conditions. These fields usually are planted early for fall grazing. Small grains can compensate for considerable injury and stand reduction, and economic injury levels have not been established. Treatment with an at-planting insecticide is justified only in heavily infested fields. Severely damaged fields may need to be replanted.

Other Information. Lesser cornstalk borers are occasional pests of >60 crops in 14 families. Grasses and legumes are preferred. Larvae can feed on decaying vegetation or weed residues buried before planting and then transfer to crop seedlings. Problems are common in summer crops that are double-cropped after winter small grains, especially when crop residues from the winter small grains have been burned.

Selected References. 57, 190

By Joseph E. Funderburk

Lesser cornstalk borer (John All, University of Georgia).

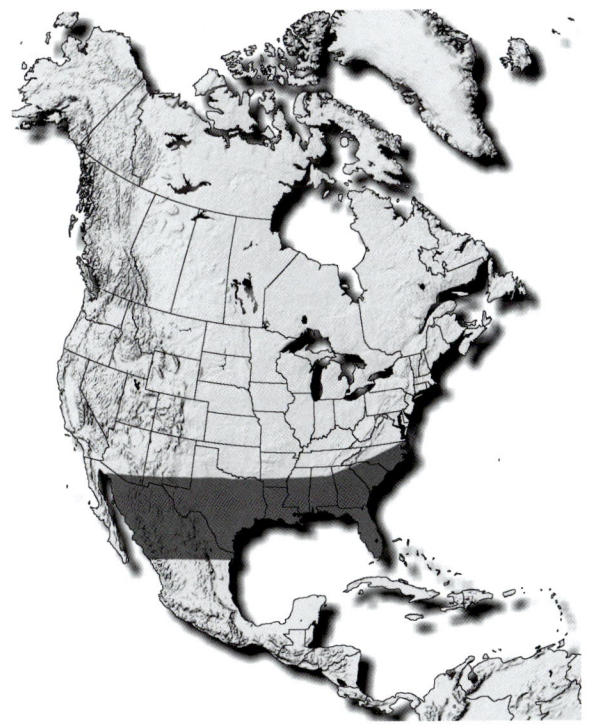
Lesser cornstalk borer distribution.

Mormon Cricket

Scientific Classification. *Anabrus simplex* Haldeman. Orthoptera (Tettigoniidae).

Origin and Distribution. Mormon crickets are native to the western Great Plains and intermountain west of the United States. Most populations are associated with low mountain ranges and sagebrush–grass plant communities. Mormon crickets have been recorded as far north and east as Manitoba and North Dakota, south to New Mexico, and west to California.

Description. Mormon cricket adults are 1 1/2–2 in. (3.8–5.1 cm) long. Their color gradually changes from tan to dark green and finally to almost black as they age. Females have a curved ovipositor measuring 1 1/4 in. (3.1

cm) long. A positive diagnostic characteristic for males is the bifurcation of the cercus.

Pest Status. Because of their large size and habit of migrating in large bands, population densities can be reach 100 crickets per square yard. Mormon crickets inspire a greater fear of damage than is normally justified. Recent studies have shown that up to 90% of Mormon cricket's diet consists of sage brush leaves and detritus. If cereal crops are in the path of a migrating band, however, extensive damage may occur.

Injury. Mormon cricket damage to wheat in the West results primarily from crickets removing and eating kernels from ripening heads. Damage seldom occurs to wheat at earlier growth stages because of the relative timing of winter wheat and cricket development.

Life History. Mormon crickets have one generation per year. During late summer and fall, gravid females deposit brown eggs into patches of bare ground where they overwinter. Oviposition may be repeated several times at several locations. Nymphs hatch in early spring. Seven nymphal instars develop over several months, particularly if cold or wet weather prevails in the spring. Elevation strongly influences cricket phenology. Therefore, life cycle development will vary within a geographic range. Mormon crickets form migrating bands when environmental conditions and food supplies favor development of large populations.

Management. Mormon crickets may exist for many years in mountain habitats at population densities of >0.1 yd^2. For reasons that are not well understood, their numbers will increase rapidly over 1–2 yr, and migrations will begin. Migrating bands can be monitored; and if wheat fields are threatened, poison bait can be placed in the path of the migrating crickets. Several commercial foliar insecticides are also efficacious.

Selected References. 113, 152

By Larry Sandvol

Mormon cricket (S. V. Romney, Utah State University).

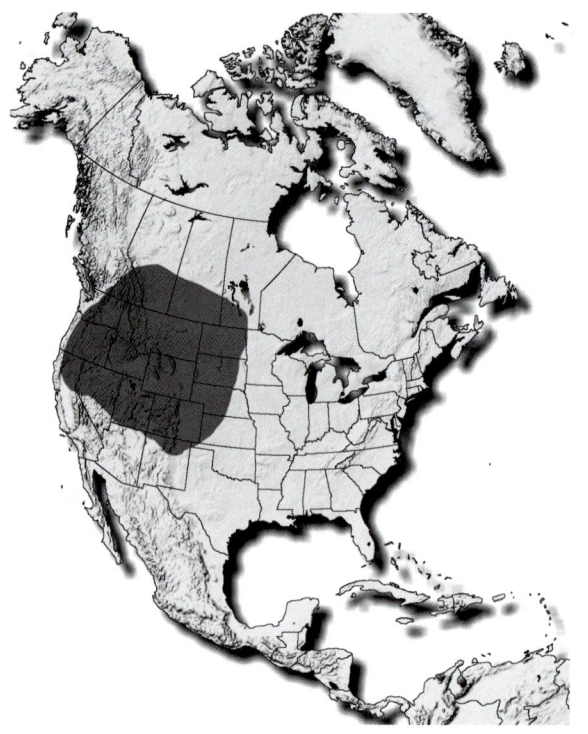

Mormon cricket distribution.

Plant Bugs

Scientific classification. Various species in the genera: *Adelphocoris, Irbisia, Labops, Leptopterna, Lopus, Lygus, Orthocephalus, Stenotus,* and *Trigonotylus*. Hemiptera–Heteroptera (Miridae).

Origin and Distribution. Plant bugs of North America comprise Old World and native species.

Pest Status. In general, plant bugs in North America have rarely caused significant damage in small grains, but some species could become troublesome under certain conditions. Of the genera named here, all contain species that feed on graminaceous hosts. *Adelphocoris rapidus* (Say) is widespread in North America; although found chiefly on legumes where it is considered a seed pest, it can occur on grains. *Irbisia* and *Labops* spp. are widespread in western North America; and at times, they are pestiferous on crested wheat grasses (*Agropyron* spp.), which are planted for range improvement. *Irbisia pacifica* (Uhler), a very common western species, has been reported to damage wheat in Washington. *Labops hesperius* Uhler may occur on wheat, but not as a serious pest. *Leptopterna dolobrata* (L.), which has been reported to attack wheat in Europe, is now distributed in North America (widely in the East, sparsely in the West). Here, it prefers timothy (*Phleum pratense*); but it

Pest Information

Adult black grass bug, *Labops* sp., and leaf feeding injury. (Jack Kelly Clark, courtesy of University of California Statewide IPM Program).

Adult Lygus bug (Keith S. Pike).

will feed on other Gramineae. *Leptopterna ferrugata* (Fallén), holarctic in distribution, feeds on grasses including wheat, but it is not known as a problem in North America.

Lygus bugs, *Lygus lineolaris* (Palisot de Beauvois) in the East and *Lygus rugulipennis* Poppius in the West, probably are the most important plant bugs attacking small grains. *Lygus rugulipennis*, recognized as a pest on assorted crops including wheat in Europe, could possibly damage grain crops in some parts of North America in the future. It is generally considered a high-latitude, high-elevation species. Several other mirids (*Orthocephalus*, *Lopus*, *Stenotus*, and *Trigonotylus* spp.), though not considered pests in North America now, possibly could become troublesome in the future.

Injury. Plant bug feeding usually occurs on reproductive parts of the plant. In some cases, sterility of the seed heads follows feeding by mirids.

Life Cycle. Plant bugs generally overwinter as eggs deposited in dried plant tissue at the base of plants. Nymphs pass through five stages. Normally, there is only 1 generation per year. Some plant bugs, such as *Stenotus* spp., will overwinter as adults.

Management. Plant bugs usually do not cause economically important damage in small grain crops and rarely require active management except in fields grown for seed production.

Selected References. 16, 75, 105, 121, 173, 174, 194, 202

By John D. Lattin

Seedcorn Maggot

Scientific Classification. *Delia platura* (Meigen), formerly *Hylemia platura* and *H. cilicrura*. Diptera (Anthomyiidae). *D. platura* and other *Delia* spp. also are called the barley shoot fly in Europe, Africa, and Asia.

Origin and Distribution. The seedcorn maggot is a native species, widely distributed across North America.

Description. Larvae, the injurious stage, are white maggots ~1/6 in. (4.2 mm) long found in the soil and on or within germinating seeds. Pupae are dark brown, barrel shaped, ~1/6 in. (4.2 mm) long and are found in the soil. Adults are small, light gray flies ~1/5 in. (5.1 mm) long. Like other anthomyiids, wings are held over the abdomen at rest giving the fly an oval appearance when viewed from the top. Many related species, such as the onion and cabbage maggots, closely resemble seedcorn maggots and require detailed examination for definite identification.

Pest Status. The seedcorn maggot is an uncommon pest of wheat. Typically, seedcorn maggot problems are associated with recently disturbed fields where organic matter, such as manure or a green cover crop (e.g., alfalfa, rye, heavy weed growth) has been incorporated. Injury also is more likely under cool, wet conditions.

Injury. Larvae feed on germinating seeds and may cause variable emergence or stand loss.

Life History. Females are highly attracted to disturbed soil with decaying organic matter, such as manure and green plant residues, and preferentially lay eggs at such sites. Generation times vary by latitude. Northern latitudes (above 43° N) have 4–5 generations occurring early spring through fall, and middle latitudes (~43–36° N) have 2–3 generations starting early spring and summer. Southern latitudes (below 36° N) have 3–4 generations occurring between fall and spring. Dormancy can begin in either the fall or summer if high temperatures are common.

Management. Seedcorn maggot problems are uncommon in wheat, and management is not warranted. With

Pest Information

Seedcorn maggots (Jim Kalisch, University of Nebraska).

serious injury, replanting may be necessary. Seedcorn maggot problems are less likely with reduced tillage and rarely occur with no-tillage.

Selected References. 69, 189

By Leon G. Higley

Stalk-Boring Caterpillars

European Corn Borer

Scientific Classification. European corn borer, *Ostrinia nubilalis* (Hübner). Lepidoptera (Crambidae).

Origin and Distribution. European corn borer, introduced into the United States before 1917, is found throughout the United States and southern Canada east of the Rocky Mountains.

Description. European corn borer moths are buff colored and have wingspans that vary from 7/8 to 1-1/4 in. (2.2–3.1 cm). Females are larger and lighter colored with fewer inconspicuous spots on the wings than males. Eggs are laid in masses of 5–50 on leaves. The overlapping eggs are initially white, then turn yellow and finally black as they develop. A newly hatched larva has a black head and pale yellow body ~1/8 in. (3.2 mm) long. The terminal fifth instar is ~1 in. (2.5 cm) long and gray to light brown. Larvae have two light brown spots on top of each abdominal segment. The pupa is reddish brown with a pointed abdomen.

Pest Status. European corn borer is a general feeder on >200 host plants. It is a major pest of corn but is usually not a serious pest of small grains. Often small grains with lush foliage, especially wheat, oats, and barley, provide suitable environments for moth aggregation. Aggregation sites give moths shelter during the day and a place to mate at night. Moths will lay eggs in these sites if preferred hosts are not available. This usually occurs in late spring before corn is attractive (before 6th-leaf stage). Small grain infestations also occur when European corn borer populations are extremely large. Infestations have been noticed in winter wheat from New York to Georgia and in oats in Iowa. Severe infestations in winter wheat (40% of stems) have occurred in New York.

Injury. Small larvae of the European corn borer feed on the leaf surface, usually in the plant whorl, then tunnel into the stalk. This results in a "shot hole" appearance as the leaves unfold. Usually small grain plants are

European corn borer larvae (Marlin E. Rice, Iowa State University).

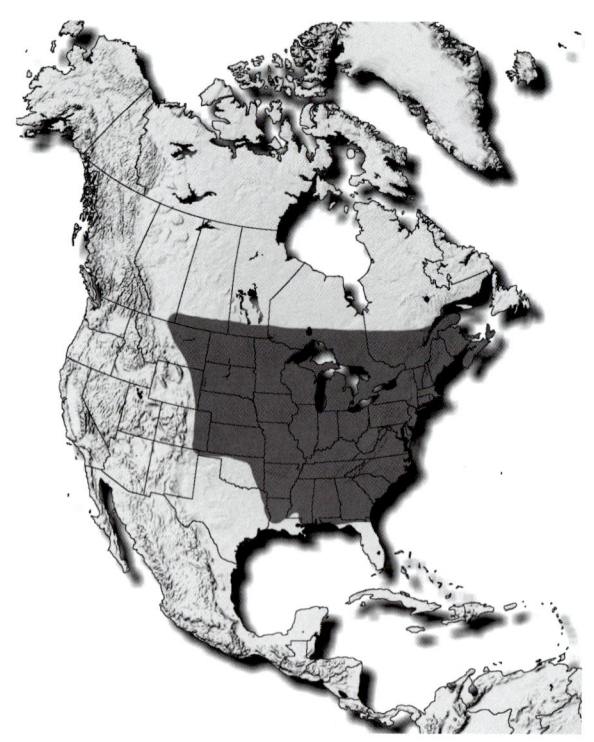
European corn borer distribution.

not large enough to sustain complete larval development, so larvae commonly move among plants. Larger larvae burrow directly into the stalk and then tunnel inside the stalk. Frass and silk are often found near entrance holes. Often corn borer injury causes plant heads to senesce prematurely. These "white heads" break easily when pulled. Holes weaken the stalk and often cause it to lodge or break.

Life History. Mature European corn borer larvae overwinter in corn stalks and ears, weeds or plant debris. They pupate, emerge as moths, and mate in the late spring. Females lay masses of eggs usually on the underside of leaves. Each female deposits ~300–500 eggs during her lifetime. Eggs hatch depending on temperature in ~3–7 d. Larvae develop through five stages in 25–35 d. There are 1–4 generations of corn borer in North America. Several ecotypes are based on the number of generations per year and different pheromone sensitivities.

Management. Producers do not use any management tactics to control European corn borer on small grains. No economic thresholds for eggs and larvae have been established for any of the small grains.

Stalk Borer

Scientific Classification. *Papaipema nebris* (Guenée). Lepidoptera (Noctuidae).

Origin and Distribution. The stalk borer is native to North America and is found throughout the United States and southern Canada east of the Rocky Mountains.

Description. Stalk borer moths are light gray with distinct white spots and have wingspans that vary from 1 to 1-1/2 in. (2.5–3.8 cm). The young larva has brown or purple longitudinal stripes that combine to form a saddle-like marking behind the true legs. This marking fades as the larva ages. The mature larva is 1-1/6–2 in. (3.0–5.0 cm) long.

Pest Status. Stalk borers are general feeders that will bore into almost any soft-stemmed plant that is large enough to hold its body. They are only incidental pests of small grains. Highest populations occur near field edges that have large-stemmed weeds, particularly giant ragweed. Typically more stalk borers are found in no-till plantings.

Injury. Stalk borers injure plants by burrowing into the base of the stem and tunneling up. This causes leaves to wilt and die, and often stunts the plant. Stalk borers also injure plants by entering through the whorl and tunneling down. In this case, leaves appear to be "chewed-up" or ragged as they unfurl. Both forms of injury can result

Stalk borer (Marlin E. Rice, Iowa State University).

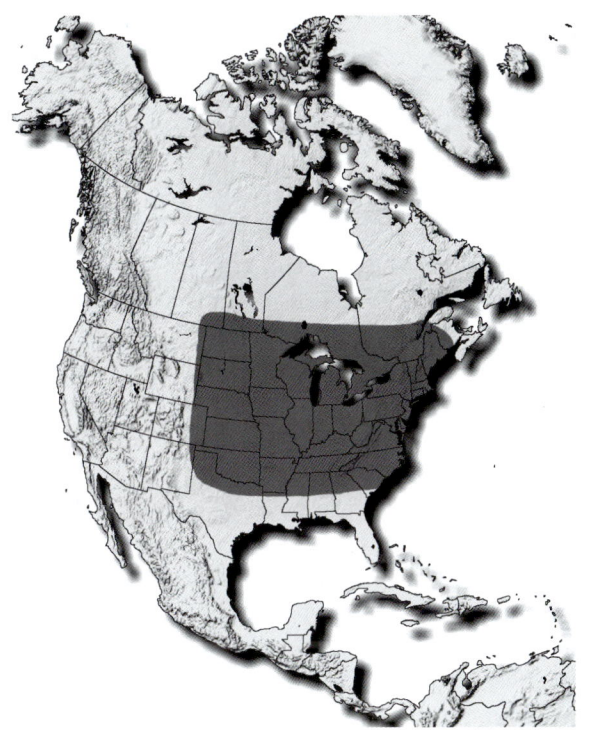

Stalk borer distribution.

in lodging, breakage, or death. Sterile heads that senesce prematurely also occur depending on the stage of plant development. Stalk borer larvae move among plants, usually from smaller grasslike plants to larger plants. Small grain plants usually are transitional hosts because they are not large enough to sustain large stalk borer larvae.

Life History. Stalk borers overwinter as eggs on weedy plants. Larvae hatch in the spring and tunnel into grass or small weed stems. They feed in the stems until they kill or outgrow the host. Larvae then migrate to a plant with a larger stem. Stalk borers develop through 7 or more stages until they pupate midsummer for 4 wk. Moths emerge late summer and deposit eggs singly or in masses. There is 1 generation per year.

Management. Producers do not use any management

Pest Information

tactics to control stalk borer on small grains. No economic thresholds for eggs and larvae have been established.

Selected References. 20, 116, 184, 205

By Richard L. Hellmich

Stink Bugs

Scientific Classification. Southern green stink bug, *Nezara viridula* (L.); rice stink bug, *Oebalus pugnax* (F.); and brown stink bug, *Euschistus servus* (Say). Hemiptera–Heteroptera (Pentatomidae).

Origin and Distribution. The southern green stink bug is found throughout the southern United States, south of a line from Virginia to Oklahoma. Recently, this pest has become established in parts of California. The rice stink bug has a broad range throughout the United States, from New York south to Florida and west to Nebraska, Colorado, and Arizona. The brown stink bug is also widely distributed throughout the United States; it is found everywhere except in northwestern states.

Description. The adults and immature stages of these three species injure wheat, oats, barley, and other grain crops. All stink bugs lay their eggs in masses and complete five nymphal stages. The eggs of the southern green and brown stink bug are arranged in an organized pattern with several rows and columns of ~40 eggs per mass. The rice stink bug eggs are arranged along two rows and also contain ~40 eggs per mass. The first instars cluster on the egg mass and apparently do not feed on plant tissue. The second instars remain congregated near the egg mass and begin feeding on the surrounding leaf and developing seed tissue. The third through fifth instars grow with each molt, disperse from the egg mass, and feed primarily on the developing grain seeds. The third instars tend to remain clustered as they move about the plant, and the last two instars lack the clustering behavior. The fifth instars of these species have distinct wing pads extending from their shoulders. Adult southern green stink bugs are pale green, have reddish bands on their antennal segments, and are 5/8 in. (15.9 mm) long and 3/8 in. (12.2 mm) wide. Adult rice stink bugs are light brown, have sharp pointed shoulders; they are small and narrow, 3/8 in. (12.2 mm) long and 1/8 in. (3.2 mm) wide. Brown stink bug adults are brown, have smooth shoulders, are more rounded in appearance and 1/2 in. (12.7 mm) long and 3/8 in. (12.2 mm) wide.

Pest Status. Stink bugs are considered a minor pest of small grain crops. However, serious crop injury can occur in isolated fields, particularly in the southern United States where stink bug populations are generally much higher.

Injury. Nymphs and adults of these stink bug species

Brown stink bug nymphs and adult (Jeremy K. Greene).

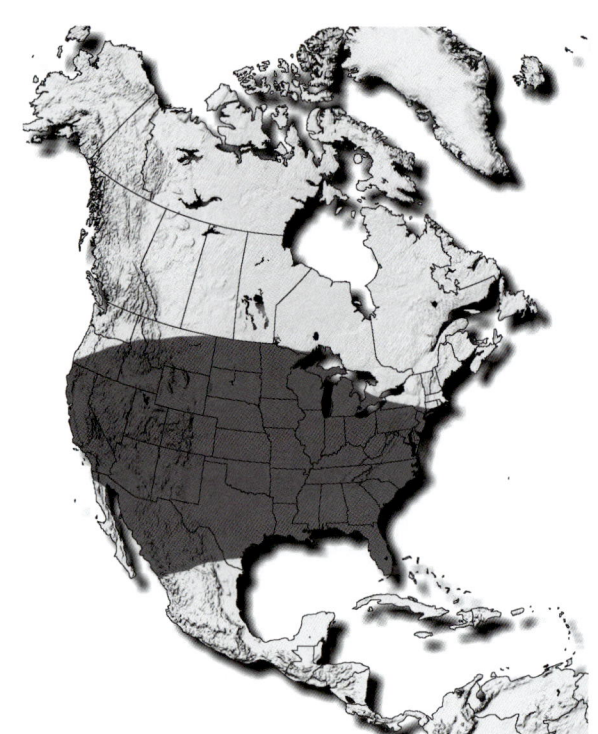

Brown stink bug distribution.

pierce and suck plant fluids from wheat and other small grain crops. Small grains are not susceptible to stink bug damage until the grains develop. Stink bugs attack the developing seeds, specifically feeding on the kernels during the milk, soft-dough, and hard-dough stages of development. Feeding during the milk stage decreases seed germination, kernel weight, and baking quality. Kernels injured during the dough stages have less damage.

Life History. The three stink bug pests overwinter as adults in protected areas such as leaf litter, wood piles, and coverings over shelters. Adults become active during warmer weather, feeding on numerous wild host plants

Pest Information

Southern green stink bug eggs, nymphs and adult (Jeremy K. Greene).

Rice stink bug adult (Jeremy K. Greene).

Southern green stink bug distribution.

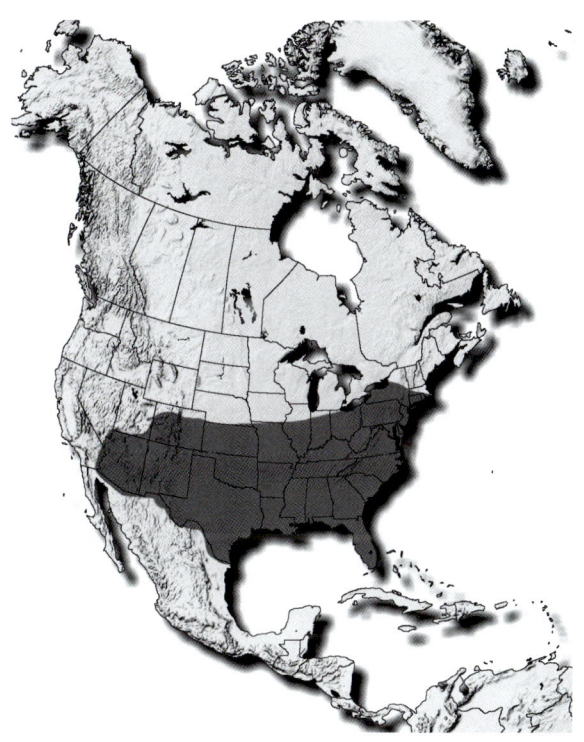

Rice stink bug distribution.

and early season crops. Mating begins in early spring, and eggs are laid on these alternative host plants. Later in the season, eggs are laid on small grain plants with developing seed heads. Eggs hatch in ~1 wk, and the nymphs complete their development in 3–4 wk. After the stink bugs complete 1 generation on the small grain crop, the new generation adults migrate to adjacent fields where more suitable host plants (plants in their early seed development stage) are available.

Management. Direct observation of grain heads, as well as use of ground cloth, beat-bucket, and sweep net, can be used to monitor stink bug populations on small grain crops. Control measures are reportedly not warranted except during the milk stage of development, and then only when populations reach 2 bugs per 20 heads. Several egg, nymph, and adult parasitoids, and numerous arthropod predators attack stink bug pests and help maintain population densities below economic injury levels.

Other Information. Although stink bugs are considered a minor pest on small grain crops, these crops serve as a bridging host for subsequent generations of these pests on summer crops. Stink bug damage to soybeans, cotton, and numerous vegetable and nut crops can be severe in some years. The late-season population buildup of stink

Pest Information

bugs on these crops and the resultant crop injury are dependent on stink bug survival and reproduction on the spring host plants such as wheat.

Selected References. 23, 119, 192, 195

By Robert M. McPherson and Jeremy K. Greene

Thrips

Scientific classification. Grain thrips, *Limothrips cerealium* (Haliday); barley thrips, *Limothrips denticornis* Haliday. Thysanoptera (Thripidae).

Origin and Distribution. Both species are native to Europe, parts of Asia, and Siberia. Barley thrips was first found in the United States in 1923 and in the northern Great Plains in 1946. Currently, the range of the barley thrips extends west from northeastern North America to southern Idaho and is rarely found south of northern Illinois. Grain thrips is now considered cosmopolitan in distribution.

Description. The adult barley thrips is <1/12 in. (2.1 mm) long and dark brown to black. On the third antennal segment, there is a toothlike projection. Nymphs range in color from white to light green. The adult grain thrips is ~1/16–1/12 in. (1.6-2.1 mm) and dark brown. Unlike barley thrips, the toothlike projection on the third antennal segment of the grain thrips is small. In both species, only females are winged.

Pest Status. Insecticides applied for barley thrips control have resulted in an increase in barley grain quantity and quality. Some data indicate that grain thrips increase pollination of small grains, thus increasing yield. There are, however, other reports that injury by grain thrips results in loss of yield and seed germination. Although the barley thrips is common in barley in the Great Plains, producers rarely consider it a major pest.

Injury. Adults and nymphs injure small grains by rasping the flag leaf and sheath and imbibing plant assimilates. Leaf-feeding injury can occasionally be severe on seedlings, although the plants usually grow out of this injury. Direct feeding on reproductive tissue also may result in reduced seed weight and germination. Usually, barley thrips injury is limited to barley; and it is unusual for this species to economically affect wheat, rye, or oats. Grain thrips can be found on all types of small grains, but it is more common on wheat, rye, and oats.

Life History. For both species, the adult is the overwintering stage, usually in sheltered areas. The female barley thrips inserts up to 20 eggs into leaf sheath tissue, usually the flag leaf. Each female will deposit ~100 eggs during her lifetime. The female grain thrips deposits eggs in the glume of the developing seed. For both species, the eggs hatch in ~3-10 d, and both species progress through four developmental stages. There are two larval stages lasting ~17-30 d followed by the prepupal and pupal stage. The prepupal stage lasts only a few hours, and the pupal stage lasts between 2 and 6 d. Although both species can have 2 generations per year, the barley thrips is often restricted to a single generation because of the climatic restrictions in its geographic range in North America.

Adult thrips (North Dakota State University).

Thrips injury to flag leaves and grain heads (North Dakota State University).

Management. In most cases, producers do not use any management tactics to control the thrips on small grains. In the northern Great Plains, producers can use a sequential sampling plan for barley thrips on barley with a treatment threshold before heading of 7.5 adult thrips per stem. North American data are not available for a treatment threshold for the grain thrips, but European researchers have documented economic yield losses at a density of 20 adults per stem.

Other Information. Although not considered to be a pest of cereal grains, in the southern United States, the tobacco thrips, *Frankliniella fusca* (Hinds), can build up large populations on winter small grains during grain filling in the spring. As the crop senesces, grain and tobacco thrips disperse to adjacent fields of summer crops where they may injure seedlings.

Selected References. 8, 21, 164

By Michael J. Weiss

Wheat Jointworm

Scientific Classification. *Tetramesa tritici* (Fitch); barley jointworm, *Tetramesa hordei* (Harris). Hymenoptera (Eurytomidae).

Origin and Distribution. This native insect is found throughout wheat-growing areas east of the Mississippi River and in many western wheat production areas. Barley jointworm occurs in the northeastern U.S., Prince Edward Island, and Ontario.

Description. The adult wasps are shiny, black, and ~1/8 in. (3.2 mm) long. There are yellow markings on the legs. Larvae are yellow, grublike, and ~1/6 in. (4.2 mm) long when fully grown.

Pest Status. Jointworm currently is a minor pest of wheat. It was an important pest of wheat in the past, but there are no recent reports of widespread damaging infestations.

Injury. Larval feeding causes galling of the walls of wheat stem near the joints. Height of the galls depends on the growth stage of the wheat during oviposition. By harvest, the galls are hard and woody. Stems are weakened by the galls and may break or lodge. Kernel numbers and grain quality are reduced by infestations. Larvae overwinter within the galls. Wheat is attacked, and barley, rye and other grasses are hosts. Barley is the only known host of barley jointworm. Wheat jointworms can be distinguished from Hessian fly larvae, which also weaken stems, but do not cause stem galling.

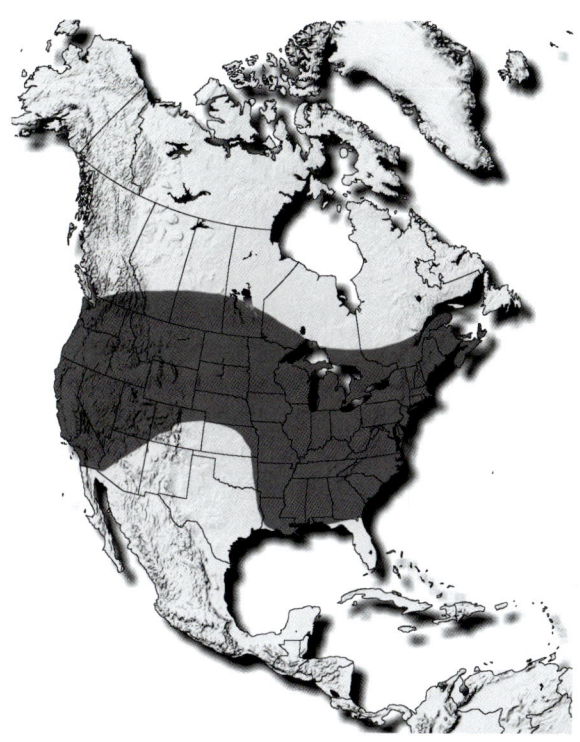

Wheat jointworm distribution.

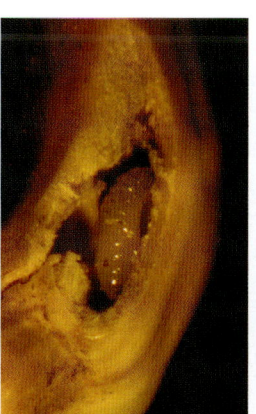 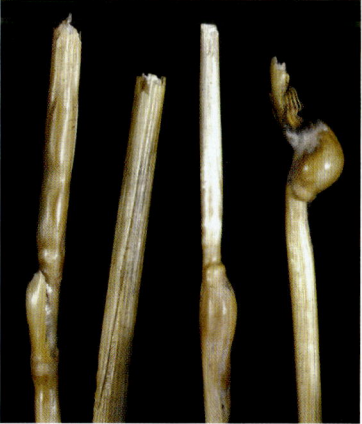

(Left) Jointworm larva in gall (Sue L. Blodgett). (Right) Galls caused by the wheat jointworm on wheat stems (Wendell L. Morrill).

Life History. Adults emerge from small circular holes chewed in the stem gall and occur when wheat is jointing. Females lay several eggs in stems just above joints. Newly emerged larvae form cells in the stem wall and feed on plant sap. Several larvae may develop within each gall.

Management. There are no current management guidelines or insecticide recommendations for this pest. Reduced tillage probably enhances the prevalence of the jointworm. Late seeding of spring wheat and crop rotation may suppress populations. The effect of parasitoids is not well known. Resistant varieties have been developed.

Selected References. 35, 95, 134, 153

By Sue L. Blodgett and Wendell L. Morrill

Wheat Midge

Scientific Classification. *Sitodiplosis mosellana* (Géhin); Diptera (Cecidomyiidae); also known as the orange wheat blossom midge.

Origin and Distribution. The wheat midge is native to Europe and was first reported in eastern Canada in 1819. The insect is now found in wheat-growing areas of Nova Scotia, Quebec, Manitoba, Saskatchewan, Alberta, and British Columbia. In the United States, wheat midge has recently been reported in North Dakota, South Dakota, Minnesota, and Michigan. Historically, populations have been reported in the northeastern United States and the

Pest Information

Pacific Northwest.

Description. The adult wheat midge is a small orange fly between 1/12 and 1/8 in. (2.1–3.2 mm) long. The adults have prominent black eyes and relatively long legs and antennae. Males have longer, more plumose antennae than females. The wings are oval-shaped, clear, and fringed with fine hairs. The larvae are 1/24–1/8 in. (1.0–3.2 mm) long and feed on the surface of developing wheat kernels. At maturity, larvae drop to the ground and overwinter in spherical cocoons at depths of ~2 in. (5 cm). Cocoons are 1/24–1/12 in. (1.0–2.1 mm) in diameter and are often coated with soil particles.

Pest Status. Wheat midge is a major pest of spring wheat in the northern Great Plains. All varieties of spring wheat are vulnerable to attack by wheat midge; however, damage varies depending on the type and variety of wheat grown. Wheat heads are most susceptible to damage when egg laying occurs during heading, Zadoks growth stages 51–59. Damage declines dramatically when egg laying occurs after the anthers are visible. Larvae feed on developing wheat kernels, causing lower grain yield, reduced grain quality, and lower grade. Wheat midge populations may exist at low levels for several years causing minor crop losses. However, under favorable conditions, populations can reach epidemic levels within 1–2 yr. Continuous wheat production on the same land increases the risk of damage. Moist conditions in May and June favor larval development; warm calm conditions increase egg-laying activity.

Wheat midge (Owen Olfert).

Injury. Eggs are laid singly or in clusters of up to four eggs on the florets of emerging wheat heads. Injury is caused by larvae feeding on the surface of developing kernels. Usually, only some of the florets on a wheat head are infested, and the level of infestation can vary from 1 to 8 or more larvae per floret. If two or more larvae develop within a floret, the kernel may abort or not fill properly. Mature kernels from infested florets are cracked, shriveled, or deformed. Small, lighter kernels are lost during harvesting operations, reducing grain yield. If one larva develops on a kernel, the surface is scarred and slightly depressed. Damage resembles drought or frost injury. Harvested damaged kernels lower grain quality including milling and baking properties.

Life History. Wheat midge overwinter as mature larvae in the soil. In the following spring, larval development depends on temperature and soil moisture. When conditions are dry during May and June, larvae remain dormant until the following year. When conditions are moist, larvae leave their cocoons and move to the soil surface to pupate. On the northern Great Plains, adults emerge over a 6-wk period beginning in late June or early July. The highest populations usually occur during the second or third week of July. Adults are relatively poor fliers and may be distributed over long distances by thermal updrafts and wind. Wheat midge adults are difficult to detect during the day because they remain within the crop canopy close to ground level where it is more humid. Females become more active in the evening. Most egg laying occurs at dusk when conditions are calm and

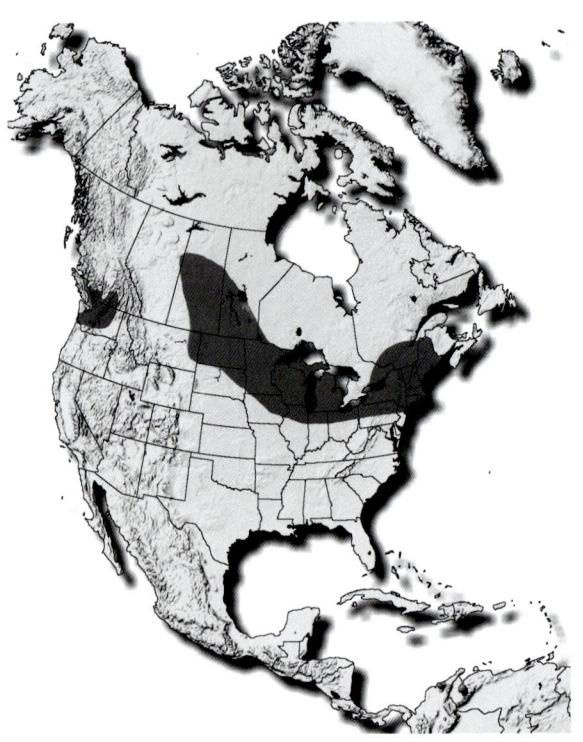
Wheat midge distribution.

temperatures are >10–11 °C. Females live 3–7 d and deposit ~80 eggs. Eggs are laid underneath the glumes or on grooves on the surface of the floret. Larvae crawl into the floret and feed on the surface of the kernel for 2–3 wk. Mature larvae remain within their cast skin in the wheat head when conditions are dry. Once moist conditions occur, larvae drop to the ground, burrow into the soil, spin a cocoon, and overwinter.

Management. Forecasts of potential wheat midge infestations can be developed from soil surveys of wheat fields after harvest. Soil samples can be evaluated to determine cocoon numbers, larval viability, and percentage of parasitism. Survey data can be mapped to identify areas at most risk to damage the following year. Significant damage and economic losses can occur when wheat midge populations in the soil reach 600 cocoons per square yard.

On the northern Great Plains, a small parasitic wasp, *Macroglenes penetrans* (Kirby) (Hymenoptera: Pteromalidae), plays a significant role in reducing wheat midge infestations. The female wasp is 1/24–1/12 in. (1.0–2.1 mm) long and lays an egg inside the egg of its host. Despite the presence of the larval parasitoid, the wheat midge larva completes its development and overwinters in the soil. The next spring, the larval parasitoid consumes its host and emerges as an adult in July. In some areas, the wasp controls 30–40% of the wheat midge population. Another parasitoid, *Playtgaster tuberosula* Kieffer (Hymenoptera: Platygastéridae), has been successfully established in western Canada to enhance the effectiveness of biological control.

Cultural practices are important in managing wheat midge. Continuous wheat cropping should be avoided to discourage the buildup of wheat midge populations. In areas where populations exceed 1,200 larvae per square yard, producers are encouraged to grow resistant crops such as canola, flax, and legumes. Other cereal crops such as barley, oats, and annual canary grass can be grown with little or no risk of wheat midge damage. Canadian varieties of hard red spring wheat, durum wheat, and soft spring wheat differ in their susceptibility to damage. With the exception of soft spring wheats, early-maturing varieties suffer less damage than late-maturing varieties. For low-to-moderate infestations, damage can be reduced by selecting less susceptible varieties, planting early, and planting seed at higher rates. These practices promote uniform heading and advance heading to avoid high populations of adult wheat midge that occur in mid- to late July. Resistant varieties may be available in the future.

An insecticide application is recommended when at least one adult midge for every 4–5 wheat heads is found at several locations in the field. Edges and center of the field should be inspected from the time wheat heads emerge from the boot until anthers are visible on most heads. Inspections should be done in the evening from 1800 hours until dusk when egg laying occurs. The timing of the insecticide application varies depending on the insecticide and its spectrum of activity against egg and adult stages of wheat midge. Guidelines from the chemical manufacturer should be consulted to ensure that the product is applied in the recommended manner.

Selected References. 47, 49, 50, 51, 142, 143, 159

By R. H. Elliott, O. Olfert and R. J. Lamb

Wheat Mites

Wheat Curl Mite

Scientific Classification. *Aceria tosichella* Keifer, formerly *Aceria tulipae, Aceria tritici, Eriophyes tulipae*. Acari (Eriophyidae).

Origin and Distribution. The wheat curl mite is widely distributed throughout the world and has a broad host range that includes several cereals and other grasses. It is primarily a pest of wheat in the Great Plains and Pacific Northwest.

Description. The wheat curl mite is very small, 1/100 in. (250 μm) long, and it has only two pairs of legs. It feeds on the leaves of wheat, corn, and other grass hosts and moves to the heads later in the season. It will almost always be found in protected areas of the plant such as a curled leaf or deep in the leaf whorl.

Pest Status. The wheat curl mite is the vector of wheat streak mosaic and high plains viruses. In the western Great Plains, wheat streak mosaic has been the most serious disease in winter wheat, but recently high plains virus has been a problem in corn and wheat. Wheat streak mosaic causes ~2% loss per year in Kansas, which is likely representative of most of the western Great Plains.

Injury. Wheat curl mite has piercing–sucking mouthparts and damages plants by sucking juices from the plants. Feeding by wheat curl mite causes the edges of young wheat leaves to remain tightly curled inward. Mites colonize this curl until the leaf is fully expanded and the curl is no longer tightly rolled. They then move back into the whorl to colonize the next developing leaf. As the plant grows, subsequent leaves or awns on the head can be trapped in the previous leaf's curl, causing distorted leaves or curled heads. Injury from mite feeding is minimal and secondary to the effects of the vectored virus. However, feeding during the heading stage, when mites can build to

Pest Information

Wheat curl mites (top) (Jim Kalisch, University of Nebraska). Wheat curl mite injury (bottom) (Gary L. Hein).

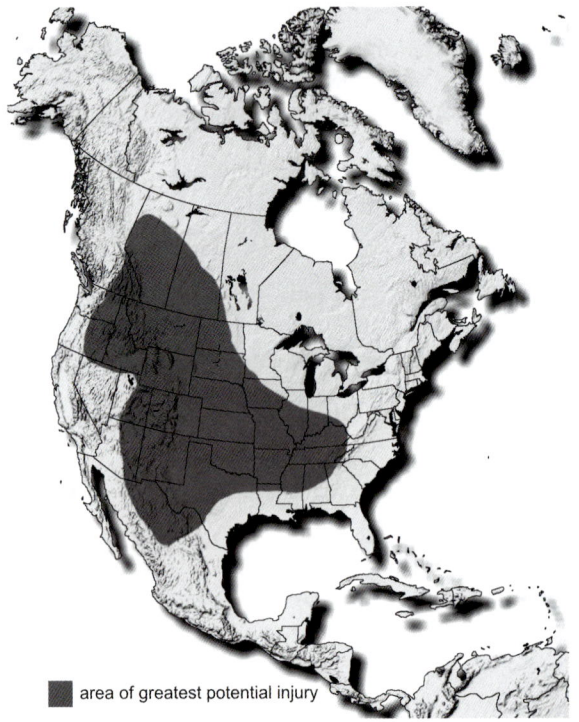

Wheat curl mite distribution.

very high populations, can affect yields by up to 10%.

Early symptoms of wheat streak mosaic are a yellow mosaic pattern of discontinuous lines on the youngest leaves. Later, more extensive yellowing, stunting, and prostrate growth will develop in severely infected plants. High plains disease symptoms typically include distinct spotting, elongate chlorotic spots, and subsequent yellowing and stunting. Plants infected early may be severely stunted and die. Wheat commonly has mixed infections; however, the combined effects of both viruses in wheat are not well understood.

Life History. Wheat curl mites have a tremendous reproductive capacity that allows them to build to very large populations when conditions are favorable. Wheat curl mites go through two nymphal stages after hatching from eggs. Development from egg to adult occurs in ~8–10 d at 25 °C. They have a continuous life cycle and overwinter in all life stages. Mites are distributed among numerous hosts via air movement.

Management. Wheat curl mites are found on winter wheat from the fall until wheat maturity the following summer. The mites cannot survive very long off green plants, so there must be "green bridge" hosts for them to survive on until new wheat is present in the fall. The most effective green bridge and greatest disease risk results when hail occurs before the wheat matures and is harvested. Kernels shelled out by the hail fall to the ground; and in the presence of high moisture, the seeds germinate rapidly and volunteer wheat begins to grow. Wheat curl mites in the heads of the wheat move to this new volunteer as the wheat matures. Infestations of mites establish colonies and transmit viruses to the volunteer. If this volunteer is not completely destroyed before the next wheat crop emerges in the fall, the mites will move from the volunteer to the new wheat crop and transmit virus.

Corn, foxtail millet, and some grassy weeds also can serve as important green bridges for the mites. The ability of these crops to serve as a bridge is determined by environmental conditions and the maturity and condition of the bridge crop when wheat is emerging. Other potential green bridges include secondary volunteer wheat, which emerges after harvest, and other grass hosts of the mites.

An important practice that reduces the potential for wheat streak mosaic and high plains disease is to avoid early planting of winter wheat. Early planting allows for a shorter green bridge period and allows a longer period of time for mites and virus to build up in the winter wheat in the fall.

Wheat varieties have been developed that show resistance to the wheat curl mite; however, mite population biotypes in some regions can overcome this resistance. There are also wheat varieties that have some resistance to wheat streak mosaic virus, but much better sources of resistance are needed. Some new, recently released wheat varieties have much greater levels of resistance to wheat streak mosaic virus. As yet, there is no known resistance to high plains virus in wheat. Chemical control of the wheat curl mite has limited effectiveness.

Brown Wheat Mite

Scientific Classification. *Petrobia latens* (Müller). Acari (Tetranychidae).

Origin and Distribution. The brown wheat mite has a worldwide distribution. It occurs wherever small grains are grown, but it is most commonly associated with wheat in the low-moisture wheat-producing areas of the western United States.

Description. Brown wheat mites are metallic dark brown to black with pale yellow legs. They are ~1/50 in. (0.5 mm) long, and their front legs are at least twice the length of the remaining three pairs.

Pest Status. Brown wheat mite sporadically damages wheat and is associated with plants that are under drought stress. It feeds primarily on cereals and grasses, but will feed on some broadleaf plants as well. Brown wheat mite also transmits barley yellow streak mosaic virus in the northern Great Plains (Montana).

Injury. Brown wheat mite has piercing–sucking mouthparts and damages wheat by sap feeding. Brown wheat mite feeds at the leaf tips causing a stippling or yellowed appearance of the leaves. Further damage results in bronzed or brown plants. This scorched appearance is similar to drought stress.

Life History. Brown wheat mite is parthenogenic and over-summers as a dormant white egg. Eggs hatch under moist conditions in the fall; but mite population increases are greatest under dry conditions. There are multiple generations from fall through spring with a generation time of about 21 d. Eggs laid during this time are red and hatch in ~7 d at 22 °C. Populations peak in early spring and decline with the onset of continuously warm weather. Females of the last spring generation lay white eggs in the soil, but these eggs do not hatch until fall.

Management. Brown wheat mite is often a problem under continuous wheat culture or where volunteer wheat was present the previous spring. Crop rotation reduces the potential for heavy infestations. Mites are most likely seen on the foliage in the early afternoon of warm days. The decision to control brown wheat mites is difficult because

Brown wheat mite (top) (Phil Sloderbeck). Brown wheat mite injury (bottom) (Frank B. Peairs).

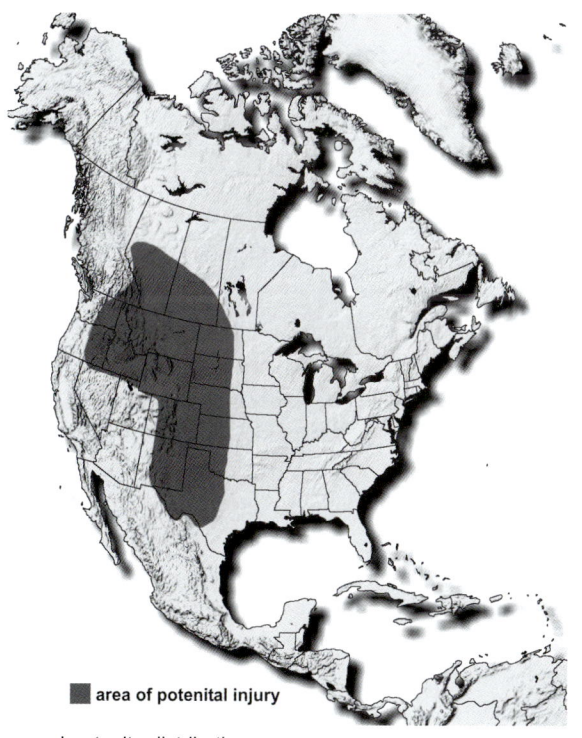

Brown wheat mite distribution.

Pest Information

severe drought stress is usually involved. If control action is taken and no rainfall is received, yield will likely be low. If rainfall is received, control action will not be necessary because rainfall of >1/4–1/2 in. (6.4–12.7 mm) usually reduces mite populations and reduces plant stress. An acaricide treatment should only be considered if there are >200 mites per row in. (50 mites/tiller); damage symptoms are evident; and females are still depositing primarily red (active) eggs. Increased presence of white (dormant) eggs indicates active populations are declining.

Banks Grass Mite

Scientific Classification. *Oligonychus pratensis* (Banks). Acari (Tetranychidae).

Origin and Distribution. The Banks grass mite has primarily a New World distribution and is associated with low moisture conditions. This mite is most commonly associated with wheat in the low moisture areas of the western United States.

Description. Banks grass mites may be colorless; but adults, when feeding, have a deep green color in the posterior two-thirds of the body. Overwintering females will be bright orange.

Pest Status. Banks grass mite is a major pest of corn in the Great Plains, and it will also feed on cereals and sorghum. It sporadically damages wheat and is associated with drought-stressed plants. Banks grass mite can damage winter wheat in the fall because overwintering females feed on the crown of the plants. Damage can build to severe levels in the spring if wheat is under drought stress.

Injury. Banks grass mite has piercing–sucking mouthparts and damages wheat by sap feeding. Banks grass mite also causes leaf stippling and yellowing, and severely damaged plants become brownish yellow. Infested leaves are heavily covered with webbing.

Life History. Banks grass mite populations can be high in the fall in corn fields. Orange, overwintering females move to adjacent wheat or other grass hosts and feed on the crown through the fall and winter. In the spring, females move up the plant to feed on leaves and lay eggs. Populations build through the spring if conditions are dry; as plants mature, mites move to alternative hosts adjacent to the wheat.

Management. Banks grass mite is predominately a problem where it may move out of adjoining corn or sorghum fields. Severe damage is uncommon and occurs in combination with dry or poor growing conditions. An acaricide application to the border of wheat fields adjoin-

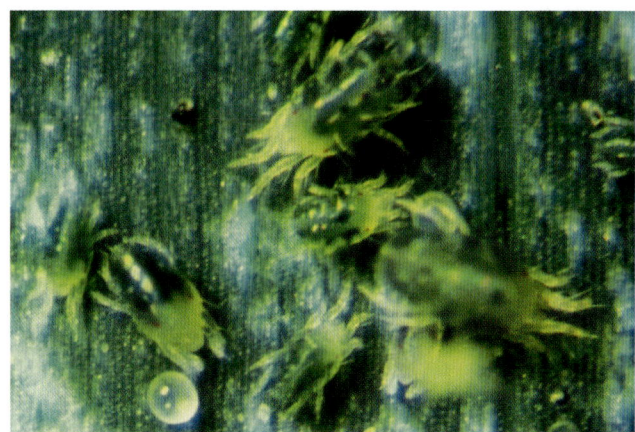
Banks grass mites (Jim Kalisch, University of Nebraska).

Banks grass mite injury (Frank B. Peairs).

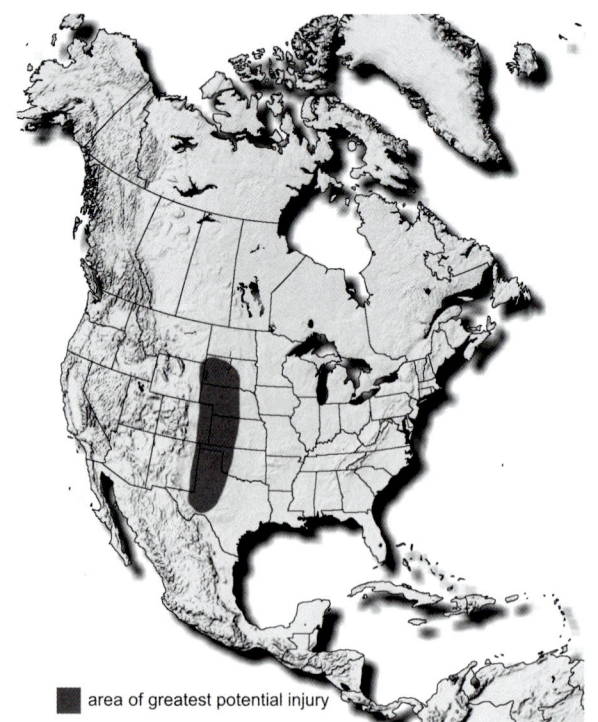
■ area of greatest potential injury

Banks grass mite distribution.

ing previously infested field corn may be warranted for severe infestations of Banks grass mite.

Winter Grain Mite

Scientific Classification. *Penthaleus major* (Dugès). Acari (Penthaleidae). Also known as blue oat mite or pea mite.

Origin and Distribution. The winter grain mite is distributed throughout the northern and southern temperate zones (25–50° latitude). In North America, damaging populations are most prevalent from south-central Kansas through central Texas.

Description. Winter grain mites are 1/32–1/16 in. (0.8–1.6 mm) long and have dark brown bodies with reddish-orange legs. Its front legs are slightly longer than the other three pairs of legs.

Pest Status. Damage from winter grain mite is most prevalent from south-central Kansas through central Texas. It prefers small grain and grass hosts, but also feeds on many broadleaf hosts.

Injury. Winter grain mite has piercing–sucking mouthparts. Sap-feeding by this mite causes the leaves to become silvery in appearance, and continued feeding results in brown leaf tips and stunted plants. Damage from winter grain mite is greatest during winter and early spring.

Life History. Winter grain mite is dormant through the summer as eggs, which are present on plants and in the soil. Egg hatch is determined by soil moisture conditions. Two generations occur annually with the first in late fall or early winter (December–January). The second generation occurs in late winter or early spring (March–April). Eggs from the second generation are dormant until fall. In the spring, mite activity is reduced during periods of warm dry weather.

Management. Winter grain mite is most severe in areas where small grains are grown continuously. Crop rotation reduces its severity. When present, the mite is difficult to sample because they feed at night or on cloudy days. High numbers of mites that cause leaf damage and stunting may warrant an acaricide application to wheat.

Selected References. 17, 72, 89, 204

By Gary L. Hein

Winter grain mite (Jim Kalisch, University of Nebraska).

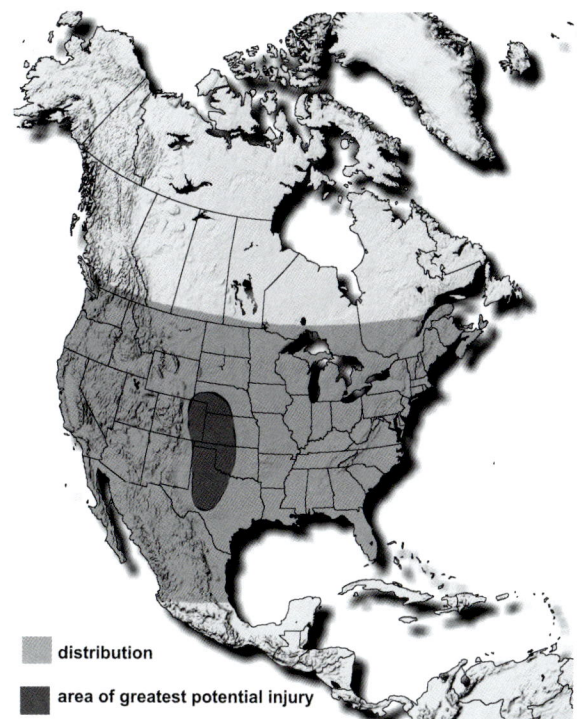

Winter grain mite distribution.

Wheat Stem Maggot

Scientific Classification. *Meromyza americana* Fitch. Diptera (Chloropidae).

Origin and Distribution. This native insect occurs in grasses across North America.

Description. There are four life stages. Eggs are elongate and white. Larvae, or maggots, are pale green and reach a length of ~1/16 in. (1.6 cm). Larvae have two black hooklike mandibles, but lack legs and head capsules. The adult flies are green or yellow with prominent black, dorsal thoracic stripes. The metathoracic legs are enlarged.

Pest Status. This is a minor pest of wheat, barley, oats, and wild grasses.

Injury. The most commonly recognized symptom of infestation is the appearance of white wheat heads that have been killed and are bleached by the sun. The stems that support the heads are severed when the newly emerged maggots move down between the stem and leaf sheath to the final feeding site above the top node. The feeding maggots rasp and tear the stems. The white heads do not produce grain. Damage to seedling or tillering plants is less obvious, although stems or tillers may be killed.

Heads and upper stems of infested plants become white, but the rest of the plant remains green until senescence. The infested heads and stems can be pulled easily from the plants, and the chewed ends of the stems are then visible. The dried stems shrink and separate from the surrounding flag leaf sheath, permitting the newly eclosed fly to escape from the plant. The white heads contrast markedly with the normal green heads, therefore casual visual inspections easily overestimate infestation levels, which are usually 1% or less. Hail, frost, plant pathogens, and other stem-boring insects also cause white heads in wheat. Accurate estimates of infestations require plant dissection to verify the presence of maggots, characteristic feeding damage, or both.

Life History. Eggs are laid on stems or leaves of young plants and on flag leaves of older plants. Larvae feed within stems. Pupation occurs between the stem and leaf sheath. After eclosion, flies exit by pushing upward from the pupation site. Most likely there are 2 generations in the northern United States and 3 generations in the South. This insect overwinters as maggots in winter small grains and wild grasses.

Management. Few management practices are recommended specifically for wheat stem maggot. Presumably crop rotation and control of grassy weeds and volunteer small grains would help reduce infestations. Several parasitoids are suspected of suppressing populations of this pest; *Bracon meromyzae* Gahan and *Coelinidea meromyzae* (Forbes) are the most common.

Selected References. 4, 61, 92, 93, 135

Wendell L. Morrill

Wheat Stem Sawfly

Scientific Classification. *Cephus cinctus* Norton. Hymenoptera (Cephidae). Related species: European wheat stem sawfly, *C. pygmaeus* (L.); black grain stem sawfly, *Trachelus tabidus* (F.). Hymenoptera (Cephidae).

Origin and Distribution. The wheat stem sawfly was first found in wild grasses in northwestern states. It currently occurs throughout the Great Plains and has been collected in other regions. The European wheat stem sawfly and black grain stem sawfly were introduced from Europe in the late 1800s and now occur in northeastern North America.

Description. Wheat stem sawfly adults are black with yellow stripes. Eggs are white and pinhead sized. Larvae are legless and white, and they are ~1/2–3/4 in. (12.7–19.1 mm) long when fully grown. Larvae will assume an S-shape when extracted from the stem. Pupae are cream-colored and show the adult colors as they mature.

Pest Status. Wheat stem sawfly is a major pest in dryland wheat grown in Montana, North Dakota, Alberta, and Saskatchewan. Some damage also occurs in Idaho, Manitoba, Wyoming, South Dakota, and Nebraska. Barley also is infested, but wild and domestic oats are resistant. Historically, only spring wheat was heavily damaged, but adults are emerging earlier and losses now

Wheat stem maggot adult (Wendell L. Morrill).

White wheat head killed by a wheat stem maggot (left) and non-infested head (Wendell L. Morrill).

Wheat stem maggot feeding above the top node of a wheat stem (Wendell L. Morrill).

Pest Information

Wheat stem sawfly ovipositing in a wheat stem (Wendell L. Morrill).

Wheat stem sawfly larva and infested stem (Wendell L. Morrill).

Wheat stem sawfly pupa and stem "stub" (Wendell L. Morrill).

occur in winter wheat. In some fields, 100% of the head-bearing stems are infested. There are no reports of significant economic losses caused by the European wheat stem sawfly.

Injury. Sawfly damage is the result of larval feeding and boring in stems. Yields of infested stems are reduced by ~20% because plant sap cannot pass up through stems to the developing grain. Females prefer to oviposit in the largest stems, which have the potential to produce the most grain. The effect of infestations on grain protein content is unclear. Sawflies do not cause white stems or premature senescence in wheat. Darkened stem areas may appear below damaged nodes. Sawfly infestations also cause harvest problems. Larvae cut notches around the interior circumference of lower stems. These weakened stems may lodge. Some of the heads that fall to the ground are not recovered during harvest because combines would have to be operated more slowly and closer to the ground, extending harvest periods and increasing equipment damage from rocks.

Wheat lodging caused by wheat stem sawfly injury (Wendell L. Morrill).

Life History. Larvae overwinter in "stubs," which are the lower sections of cut stems. The obligatory diapause is completed after at least 90 d of low temperature. Pupation in spring is completed in ~2 wk, although the duration is affected by temperature. Adults emerge by pushing through soft plugs in the upper ends of the stubs. Females appear 1–2 d earlier than males. Females mate once, but males may mate several times. Females are attracted to volatile chemicals that are produced by green wheat plants. They fly upwind and begin to oviposit as a suitable host is encountered. Flights are short and occur during calm sunny periods. Adults lay up to 50 eggs and live for ~5 d. Eggs are inserted into stems. Stems may receive eggs from several females. Egg fertilization is controlled by the female at the time of oviposition; fertilized eggs produce female offspring and unfertilized eggs produce males. Fertilized eggs are more likely to be laid in the largest and

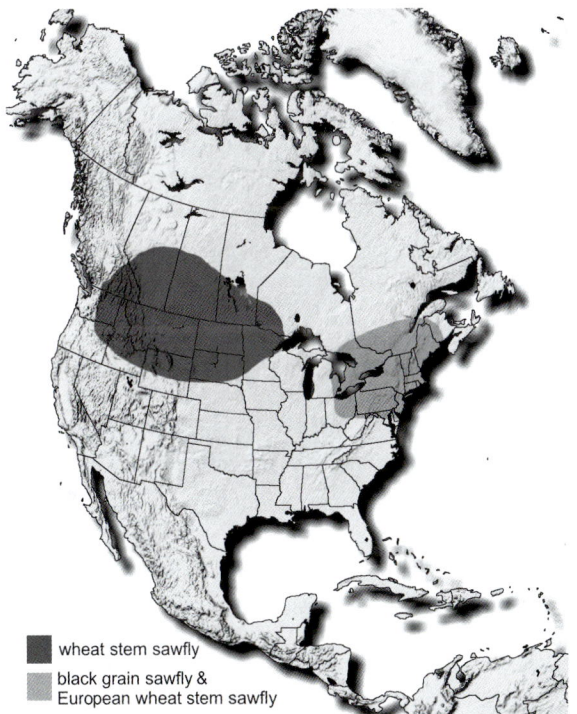
Wheat stem sawfly, black grain sawfly and European wheat stem sawfly distribution.

Pest Information

most suitable stems. Females from the largest stems are larger, live longer, and produce more eggs.

Larvae are cannibalistic, and only one survives. Wheat is susceptible to attack during stem elongation through flowering, after which larvae do not have time to complete development before plant senescence. Larvae bore throughout stems. Infested stems are filled with frass. Larval infestation levels can be estimated by examining mature stems. Visual estimation of lodging is not an accurate method of determining larval infestation because not all infested stems lodge, and parasitized larvae do not cut stems. Other insect larvae that feed in stems and fill them with frass include stem-boring Lepidoptera and the wheat stem maggot.

Management. Sawfly adult activity can be monitored by using sweep nets and monitoring traps that contain a plant volatile chemical and insect pheromone. However, insecticide applications are not effective for adult sawfly management because adult flight activity can exceed 1 mo. Because of the short residual period of currently registered pesticides, repeated applications would be necessary to provide plant protection.

Solid-stemmed wheat cultivars have been developed. These cultivars are infested by sawflies, but larvae bore less extensively and the effect on yield is reduced. Unfortunately, solid-stemmed wheats have lower yield potential. Solid-stemmed cultivars usually only yield more than hollow-stemmed varieties under intense sawfly pressure.

Late-planted spring wheat avoids attack if stem elongation begins after the annual sawfly flight season ends. In experimental trials, trap strips of winter wheat have been effective in intercepting adult flights into spring wheat fields. Tillage has minimal effect on adult emergence. Some growers swath wheat to prevent lodging losses. Field edges are more heavily infested; hence, only the borders may need to be swathed. Swathing has no effect on larval survival.

Sawfly larvae are attacked by two species of native, host-specific braconid parasitoids, *Bracon cephi* (Gahan) and *B. lissogaster* Muesebeck. Larvae are paralyzed at the time of attack by the parasitoid. Thus, feeding terminates and stems are not cut. The current trend toward reduced tillage has increased the activity of parasitoids. Parasitism exceeds 95% in some fields; and after several years, the parasitoids seem to cause a decline in the sawfly population. Sawflies rapidly adapted from wild grasses to wheat, but parasitoids have been slower to adapt to wheat.

Selected References. 136, 138, 139, 172

By Wendell L. Morrill and Michael J. Weiss

Wheat Strawworm

Scientific Classification. Wheat strawworm, *Tetramesa grandis* (Riley). Hymenoptera (Eurytomidae).

Origin and Distribution. Wheat strawworm occurs primarily in wheat-growing areas west of the Mississippi River and is less common in eastern North America.

Description. First-generation adults are black, wingless, antlike insects ~1/10 in. (2.8 mm) long. Second-generation adults are winged and about twice as large. Larvae reach ~1/4 in. (6.4 mm) in length and are yellow green.

Pest Status. Strawworm is a minor pest of wheat. Strawworm was an important pest of wheat in the Great Plains in the past, but there are no recently reported cases of outbreaks.

Injury. First-generation larvae stunt young plant growth and typically kill the plant or damage the crown thereby preventing grain head development. Second-generation larvae tunnel in the stems, causing stunting, straw weakness, and reduced grain development.

Life History. Wheat strawworm only attacks wheat. There are 2 generations per year. Larvae and pupae overwinter in stubble or straw. Adults emerge in March or April and oviposit in young wheat plants. Larvae feed within the plants and pupate; adults emerge in May. The second-generation adults are females and reproduce parthenogenetically. Eggs are inserted into stems, and the resulting larvae feed in upper regions of the host stems. Pupation occurs at

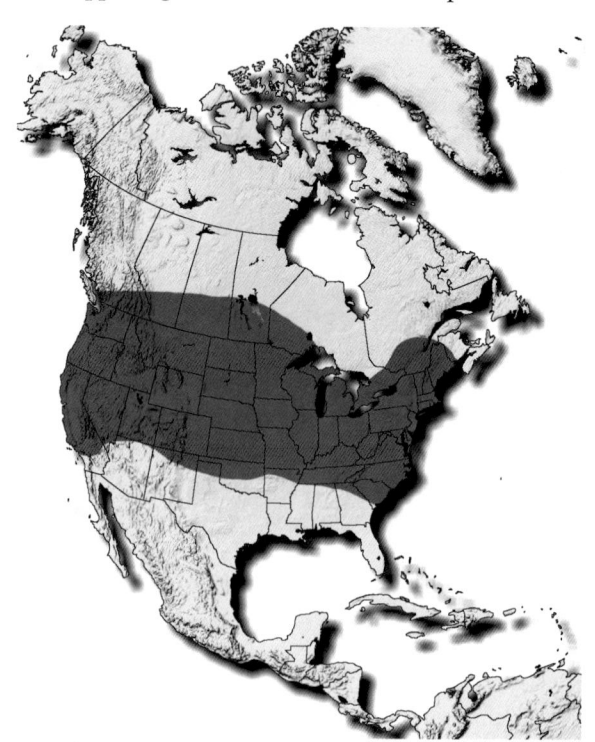

Wheat strawworm distribution.

Pest Information

(Top) Wheat strawworm adult (Wendell L. Morrill). (Bottom) Wheat strawworm larva and damaged stem (Wendell L. Morrill).

the feeding site.

Management. Estimates of 22% grain loss have been attributed to wheat strawworm. Strawworm infestations are typically in wheat adjacent to infested stubble from the previous year. The wingless first-generation adults have limited dispersal, therefore crop rotation and isolation of wheat fields are suggested. Destruction of volunteer host plants may be useful. Wheat strawworm is attacked by several species of parasitoids.

Selected References. 94, 96, 123, 154

By Sue L. Blodgett and Wendell L. Morrill

White Grubs

Scientific Classification. All larval forms of the scarab beetle family are referred to as white grubs. True white grubs, *Phyllophaga* spp., include *P. anxia* (LeConte), *P. implicita* (Horn), *P. lanceolata* (Say), *P. congrua* (LeConte), and *P. crinita* Burmeister; Coleoptera (Scarabaeidae). Adults are called May or June beetles. Annual white grubs are in the genus *Cyclocephala,* and pests include northern masked chafer, *C. borealis* Arrow, and southern masked chafer, *C. immaculata* (Olivier); Coleoptera (Scarabaeidae).

Origin and Distribution. More than 200 species of *Phyllophaga* are present in North America, occurring in the contiguous United States, southern Canada, and Mexico.

Description. Larvae are C-shaped with a white body and brown head. Newly hatched larvae are ~1.8 in. (3.2 mm) long; mature larvae are up to 1 ½ in. (3.8 cm) long. Just forward of the anal slit, white grubs have a pattern of hairs and spines known as the raster. The raster of true white grubs has a zipper-like row of stiff spines; annual white grubs have randomly arranged hairs.

Pest Status. Both groups of white grubs can be serious pests of grasses, though the true white grubs are more common pests. Grasses found in pasture, meadows, and turf are most frequently injured. Small grains seeded into sites where grubs are abundant are at risk of stand reduction.

Injury. White grubs feed by consuming roots. Most feeding takes place in the top 2 in. (5 cm) of the soil profile. First-stage grubs are small and rarely consume enough root tissue to kill plants. Second- and third-stage grubs consume large amounts of root material, inflicting serious injury to small grains in spring (true white grubs) or late summer (annual grubs). Third-stage, true white grubs, in the spring of the final year of their life cycle, feed briefly; but stand reductions are uncommon. Spring feeding by third-stage annual white grubs is short in duration, and root injury is seldom severe.

White grub larva (Phillip A. Glogoza).

Life History. The life cycle is characterized by an egg, three larval instars, a pupa, and adult. Generation time for true white grubs ranges from 1 to 5 yr. In spring and early summer, adult beetles emerge from the soil after dusk, fly to foliage of deciduous trees and shrubs to feed and mate, and return to the soil before dawn. Common host trees include willow, cottonwood, ash, elm, oak, walnut, and birch. White grubs are often associated with coarser textured soils near the adult feeding sites. Mated females burrow into the soil and lay 30–60 eggs over several weeks.

Pest Information

Eggs hatch in 2–4 wk. In late summer and fall, the grubs molt. The second instars feed until environmental cues trigger downward migration in the soil. White grubs overwinter below the frost line in earthen cells. In the spring, grubs migrate up to feed aggressively on plant roots. These grubs molt quickly to the third instar. The third-stage grub pupates in late summer of its final year. The pupal stage lasts ~2 wk. The final winter is spent as an adult in the soil. Where 2–4-yr life cycles occur, populations include multiple broods with one brood making up the largest proportion of the population. For white grubs with a 1-yr life cycle, the biology is similar except they overwinter as third instars, pupate in early summer, and emerge as adults shortly afterward.

Management. Though small grains have been rated as moderately resistant to grub feeding, economic thresholds are as low as one grub per square foot. A single white grub feeding along a drill row can cause extensive damage. Crop rotation is best used to protect small grains from grub injury. If grub populations exceed the threshold, select tap-rooted crops, such as beans, peas, sunflower, alfalfa or clover, that tolerate grub feeding better by replacing lateral roots quickly, or plant a crop such as corn or grain sorghum that may be protected from injury with an at-planting soil insecticide treatment.

Selected References. 41, 111, 167

By Phillip A. Glogoza

Click beetle (Wendell L. Morrill).

Pacific coast wireworms (Keith S. Pike).

Wireworms and False Wireworms

Wireworms

Scientific Classification. Species of economic importance in small grains include the Great Basin wireworm, *Ctenicera pruinina* (Horn); the dryland wireworm, *Ctenicera glauca* (Germar); and *Limonius* spp. Coleoptera (Elateridae).

Origin and Distribution. Wireworms are native and occur throughout North America.

Description. Larvae are orange, cylindrical, and hardened. The size of fully grown larvae varies with species. The legs and antennae are small. Adults, or "click beetles," can snap their head downward quickly as a defense or to regain an upright position. Beetles are flattened and cigar-shaped. Color is variable; although most species are black, brown, or gray.

Pest Status. Wireworms are very important pests of small grains, potatoes, and other crops. Damage is more extensive in the northwestern United States and western Canada. Wireworms are frequently undetected because they are hidden underground except for a brief period when adults mate, disperse, and lay eggs. In Montana, they must be ranked among the most injurious insect pests of wheat, for, unlike grasshoppers and cutworms, they continue at damaging levels year after year.

Injury. Wireworms have chewing-type mouthparts. The larvae feed on seed, roots, and underground stems of growing plants. Young seedlings may be killed, and feeding wounds enhance the incidence of plant–pathogenic fungi.

In heavily infested fields, wheat may grow unevenly with patches that lack vigor. Surviving plants may have small holes in leaves from feeding that occurred before the plants emerged above the soil surface. Dead plants may be found by careful inspections above and below ground.

Feeding occurs early in the spring and also in the fall, so that winter and spring wheat are attacked. During hot, dry summer periods, larvae may go deep underground. Adults also may feed on plants but do not affect yields. Symptoms of wireworm infestation are easily confused with freezing or desiccation, or "winter kill". Early season wireworm damage is similar to that caused by false wireworms and army cutworms.

Life History. Hibernating adults overwinter. Although some adults fly, dispersal is limited. Eggs are laid in the soil in the spring. Partially grown larvae can survive in summer fallow fields, even in the absence of green plant material. The larval stage can last from 1 to 9 yr, but 2–4 yr is more typical.

Management. Soil samples can be collected and examined to estimate levels of larval infestation. Using a golf hole cutter makes it possible to obtain soil cores with precise volumes. The cores are filtered through a series of screens to separate larvae from the soil.

Solar bait stations can be used to concentrate larval activity and aid in their detection. These stations consist of ~0.4 pint (0.2 L) of wheat that is buried to a depth of 5 in. (10 cm), and covered with a 5–7 in. (12.5–17.5 cm) mound of soil. The mound is covered with clear plastic. Stations are placed in the field in the fall and remain in place until spring. The soil under the plastic is comparatively warmer and moister and provides a food source for the larvae. In the spring, the grain can be excavated and examined for wireworms. Comparison of population densities under the stations and at random in nearby soil have indicated that wireworm concentration increases by ~25 times.

Solar bait station for detecting wireworms (Wendell L. Morrill).

Emergence traps are also effective in capturing adult beetles. These traps consist of a conical wire screen base that covers 7.5 square feet (0.7m^2) and an apical retaining cage. Results indicated that over 200,000 beetles per acre were produced annually in grasslands in Georgia.

Insecticide seed treatment should be used in fields where bait stations detect a large infestation, in fields having previous history of damage, and in fields that previously were in sod. Insecticide seed treatment has resulted in yield increase of nearly 30%. No effective control practices can be implemented after wheat has been planted.

Selected References. 132, 133, 191, 198

By Wendell L. Morrill

False Wireworms

Scientific Classification. Species of economic importance include the plains false wireworm, *Eleodes opacus* (Say); *Eleodes extricata* (Say), *Eleodes hispilabris* (Say), *Eleodes suturalis* (Say), and *Embaphion muricatum* (Say). Coleoptera (Tenebrionidae).

Origin and Distribution. This group of native pests occurs predominantly in semiarid regions of western North America.

Description. False wireworms have four life stages. They are the adult (beetle), egg, larva, and pupa. The beetles are shiny or dull black. Eggs are white. Larvae are elongated and yellow and have small legs. Larval size varies with species. Larvae closely resemble "true" wireworms, but adults have longer legs and antennae.

Pest Status. Plant damage is similar to that caused by wireworms. However, false wireworms are more active,

False wireworm adult (Wendell L. Morrill).

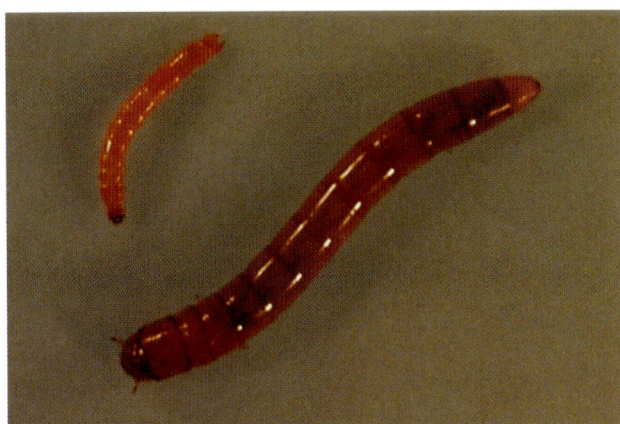
False wireworm larvae (Wendell L. Morrill, Keith S. Pike).

larger, and grow at a faster rate, and therefore are more destructive. Crops, including small grains and corn, are attacked.

Injury. Larvae feed on underground stems, roots, and seeds. Seed injury is more likely to occur during dry conditions that delay germination. Seedling and tillering plants may be killed, and one larva may kill several plants. Plants may be severed near the soil surface, thus damage is similar to that caused by cutworms. Surviving plants have a ragged appearance, and damaged stands may lack vigor. Moderate damage frequently is undetected without detailed field inspection.

Life History. Eggs may be laid throughout the small grain growing season. Either larvae or adults may overwinter. Usually there is 1 generation per year.

Management. Adults are active on the soil surface and readily captured in pitfall traps. Larvae can be found in solar bait stations as described in the section on wireworms. Larvae can be found by searching in rows where plants are missing. Where there is a history of infestation, damage may be reduced by using insecticidal seed treatments. No treatments are available for infested fields once planting has occurred. Crop rotation and summer fallow may suppress population densities.

Selected References. 5, 27, 28, 137

By Wendell L. Morrill

Insect Pests of Small Grains Outside of North America

Sunn Pest and Cereal Bugs

Scientific Classification. The name sunn pest or suni bug generally refers to *Eurygaster integriceps* Puton, but they are also commonly applied to *Eurygaster austriaca* Schrank, *Eurygaster maura* L., and other *Eurygaster* spp. that infest wheat and barley, Hemiptera–Heteroptera (Scutelleridae). Related species include the pointed wheat shield bugs, *Aelia acuminata* L., *Aelia rostrata* Boheman, and other *Aelia* spp.; and other cereal bugs, *Dolycoris pennicillatus* Horvath and *Carpocoris fuscispinus* Boheman, Hemiptera–Heteroptera (Pentatomidae).

Origin and Distribution. Sunn pests and wheat shield bugs are widespread throughout the rain-fed grain-producing regions of southern and eastern Europe, northern Africa, and southwestern and south-central Asia. The most severe outbreaks occur on the Anatolian Plateau of Asia Minor, in the northern Black Sea and Caspian Sea drainages, and in the highlands of northern Syria, Iraq,

Adult sunn pest on wheat head (ICARDA).

and Iran.

Description. Sunn pests are about the size of an adult thumbnail and range in color from light or dark brown to deep reddish brown. They have the general body shape of a stink bug, although they are generally a bit more rounded anteriorly when viewed from above. The scutellum of a sunn pest, which in most true bugs is a triangular-shaped plate in the middle of the back, extends toward the end of the abdomen like a shield. Nymphs are circular when viewed from above and are black or dark brown. Eggs are laid in clusters of 13–14 in two rows on leaves and stems of small grain plants and on broadleaf weeds. The eggs are normally pale green with red and brown markings appearing near hatching. Pointed wheat shield bugs and cereal bugs have a body shape similar to that of sunn pests, but are slightly smaller, more angular, and have a triangular scutellum.

Pest Status. The sunn pests are arguably the most important pests of wheat and barley worldwide. They are especially important pests in the dryland wheat-producing areas of the developing countries of Africa, Asia, and Eastern Europe. In these countries, sunn pest is normally the only wheat and barley pest that routinely attracts government attention in the form of subsidized regional control programs.

Injury. Wheat is the most common host; but barley, rye, oats, sorghum, and maize are also attacked. The sunn pest injures plants by sucking fluids from stems, leaves, or developing grains, and reducing plant vigor. When feeding on kernels, the sunn pest ingests the milky endosperm and can reduce kernel weight by 15–60%. Nymphs and adults inject a proteolytic enzyme while feeding that aids in dissolving plant proteins. Leaf or stem tissues surrounding the feeding site die. The enzymes remain in "stung" kernels in a dehydrated, inactive state after the insect feeds. When water is added to flour milled from infected grain, these enzymes are reactivated and destroy the dough's gluten. Such dough is runny and cannot be used to make bread.

Life History. Life histories are similar for sunn pest, pointed wheat shield bugs, and cereal bugs. The sunn pest undergoes a single generation per year. It overwinters as a sexually immature adult in mountains and hills under rocks and fallen leaves and in bark crevices. In spring, the fully mature adults migrate to cereal fields in the valleys where they begin feeding and mate and lay eggs on weeds and crop plants. The eggs hatch in ~10 d, and the neonate nymphs feed on grain plants. When sufficient body fat is accumulated in adults, the sunn pests migrate to overwintering sites. About 25% of the sunn pest population dies during a typical winter, mostly because of inadequate food reserves. The survivors begin to move into wheat and barley fields in the spring when ambient temperatures reach ~ 54 °F (12 °C), and they immediately begin to feed.

Management. Chemical sprays consisting of aerially applied, ultra-low-volume (ULV) applications of synthetic pyrethroids are the most widespread control method. There are reports that commonly used insecticides are becoming less effective because of insecticide resistance. The need for and magnitude of spray programs are determined by counting the number of overwintering sunn pests and the extent of body fat accumulation. Sunn pests with little fat body are less likely to survive. Sprays are applied to wheat fields when economic thresholds of 6–12 nymphs/m^2 or 2–3 adults/m^2 are observed. Biological control tactics include mass release of egg parasites and enhancing parasite habitat by growing shelter belts of trees around wheat fields. The use of entomopathogenic fungi to control of sunn pest has shown promise in recent years. Cultural control includes planting early-maturing wheat and barley varieties, restricting wheat and barley cultivation in high pest density areas, and disrupting the coincidence of ripe grain and young sunn pest adults by early harvesting.

Selected References. 114, 128, 130, 146, 147

Wheat Ground Beetle

Scientific Classification. *Zabrus tenebrioides* Goeze, Coleoptera (Carabidae); also called the corn ground beetle.

Origin and Distribution. Wheat ground beetle is widespread in winter and spring wheat-growing areas of western and eastern Europe, western Asia, northern Africa, and south-central Asia. It is especially serious in western Russia, Ukraine, Georgia, and Turkey.

Description. Adult wheat ground beetles are nondescript, black, and ~0.6 in. (15.2 mm) long. The elytra extend completely over the abdomen and have several indented lines running lengthwise. The tibiae and tarsi may be yellowish-red to dark black. The larvae are yellowish-white with a blackish-brown head and thorax and three pairs of short legs. Larvae may attain a length of up to 1 1/4 in. (3.2 cm).

Wheat ground beetle larvae and injured wheat seedlings (Ross H. Miller).

Pest Status. The wheat ground beetle is common throughout its range, but only occasionally causes economic crop loss in localized areas. Insecticidal control is not practiced throughout most of its range in developing countries. It is most damaging where wheat or barley are grown in monocultures without fallow.

Injury. Adult beetles feed on sown seed and on the grains of developing ears. Larvae feed on leaves and stems of seedlings and on roots. Larvae pull the leaves underground and feed on them in a subterranean chamber, leaving a bare area where the seedling once stood. Fresh deposited soil from burrowing by wheat ground beetle is usually present at the soil surface. A single larva can eat ~25 wheat seedlings. Damage is first noticed as bare spots appear along rows, with the bare space increasing daily. Infestations of up to 42 larvae per square yard (50 larvae/m^2) are common in monoculture fields and may result in complete crop loss within 7–14 d at normal spring temperatures. Non-cereal crops are rarely attacked by *Z. tenebrioides*. Damage by wireworms and scarab beetles may be mistaken for wheat ground beetle infestations. Wheat ground beetles often pull the entire seedling under ground, whereas wireworms and white grubs feed on the roots and leave the dead, yellow shoots of the seedling on the soil surface.

Pest Information

Life History. The wheat ground beetle has 1 generation per year. In western Asia, adults typically appear in May, June, and July and feed at night, hiding under field debris during the day. Mating and oviposition occur in September after aestivation during hot summer months in the lower latitudes of their range. Adults die soon after oviposition. The eggs are laid singly in the soil at a depth of 6–7 in. (15.0–17.5 cm). A single female can lay 40–80 eggs; eggs hatch within 2 wk if the soil is moist. Oviposition may be delayed until late fall under dry conditions. Larvae overwinter in the soil at depths of up to 16 in. (40 cm), and damage is usually first noticed in late February in the continental Mediterranean climate areas of western Asia and Eastern Europe. Pupation occurs in March, and adult emergence coincides with the milk stage of grain development over much of their range.

Management. Because adults and larvae feed on leaves and stems, foliar applications of insecticides are effective in controlling isolated outbreaks. Large-scale outbreaks can be almost completely eliminated by using a crop rotation program that alternates cereals with a fallow or legume cover.

Selected References. 34, 98, 127, 128

Ground Pearls

Scientific Classification. Ground pearls, *Porphyrophora tritici* Bodenheimer. Hemiptera–Sternorrhyncha (Margarodidae).

Origin and Distribution. Found throughout the wheat- and barley-growing areas of western Asia with moderate to low rainfall. Heaviest populations have been reported from central and southern Turkey and northern Syria.

Description. Eggs are very small and red and are found in the soil encased within white cottony wax filaments, hence the name "ground pearls." The first instar is small, reddish, 1/32–1/12 in. (0.8–2.1 mm) long and found between the leaf sheaths at the base of the plant. The antennae have 5 segments, and the legs terminate in a single claw. The second instar is cystlike, red or purple, legless, and 1/8–1/5 in. (3.2–5.1 mm) in diameter. It is found at the base of the plant and may or may not be covered by a leaf sheath. Adult females are oval, red, and 1/8–1/6 in. (3.2–4.2 mm). Adult males are thin, cylindrical, red, ~1/8 in. (3.2 mm) long, and winged.

Pest Status. Ground pearls most frequently attain pest status in areas planted continuously to cereal monocultures in marginally arable land. As such, ground pearl outbreaks are linked to the socio-economic condition of the farmers, who commonly plant barley monocultures

(Top) Second-instar cysts of ground pearls (Ross H. Miller). (Bottom) Ground pearl eggs deposited in soil (ICARDA).

on marginal land to feed their sheep. Ground pearls may be present, but usually are not economically serious pests in higher rainfall areas (>12 in. [30 cm] annual rainfall) or in irrigated fields. They readily attack wheat plants in the absence of barley. They may be transferred among fields on farm equipment or straw.

Injury. Ground pearls reduce the vigor of the plant by sucking plant fluids during the seedling stage of development. Infested plants are stunted and more susceptible to drought, disease, and nematodes. In severe infestations, 80–100% of plants may be infested with one or more insects, causing complete grain loss. Field studies have shown that a density of 12 insects per plant kills the plant.

Life History. Ground pearls can reproduce sexually and asexually (parthenogenesis). Eggs overwinter in cells in the soil and emerge in December in the Middle East. They immediately infest the plant after germination where they remain between the leaves and the stalk, but they sometimes tunnel to the root collar. Cysts are formed following the first molt in early April, concurrent with tillering. Adult females emerge from late May to June when heading occurs. Females remain under the soil and lay up to 300 eggs in a small chamber lined with wax threads. Males undergo a free pupal stage before becoming adults.

Management. Rotating a clean fallow or legume cover crop between cereal plantings readily controls ground pearls. Various soil insecticides are also effective, but usually are too expensive.

Selected References. 48, 127, 129

Barley Stem Gall Midge

Scientific Classification. *Mayetiola hordei* Keiffer. Diptera (Cecidomyiidae).

Origin and Distribution. The barley stem gall midge is a barley pest in regions adjacent to the Mediterranean Sea in northern Africa, southern Europe, and western Asia. Economic damage has been observed mainly in northern Africa.

Description. Adults of the barley stem gall midge resemble adults of the Hessian fly. They are gray and have a mosquito-like form. The eggs are ~1/50 in. (0.5 mm) long, slender, glossy, and pale red, becoming a deeper red near hatching. First instars are the same size and color as eggs. After 3 d, first instars turn completely white and reach ~1/16 in. (1.6 mm) before molting. Second instars initially have well-defined segmentation, but become completely smooth and cylindrical after 3–4 d. During this growth stage, plant tissue gradually forms around the larvae until it is completely embedded in a pea-sized gall. Second instars of barley stem gall midge are morphologically similar to the second instars of Hessian fly, although there are microscopic differences between them. Adults of the barley stem gall midge can be readily distinguished from Hessian fly adults by comparing male genitalia and the abdominal segments of females. The third instars and pupae of both flies are enclosed in a puparium formed from the hardened cuticle of the second instars. The puparium of the barley stem gall midge adheres tightly to the plant tissue, whereas the Hessian fly puparium readily comes loose from plant tissue.

Pest Status. *Mayetiola hordei* reduces grain yields of barley in Morocco by up to 45% during heavy infestations. Damaging infestations have been observed on barley in Algeria, Tunisia, and Libya.

Injury. The first and second instars are the feeding stages of barley stem gall midge. Eggs are laid mostly on the upper leaf surface. After hatching, first instars crawl to the base of the stem and feed under the leaf sheath. Susceptible plants are stunted and dark green. The mechanism of larval feeding is not fully understood, but it is probably similar to that described for the Hessian fly on wheat. Unlike Hessian fly, larval feeding causes a distinct gall to form at the feeding site. Puparia are always embedded in galls.

Barley stem gall midge puparia in gall formed at the base of a barley shoot (Ross H. Miller).

Life History. The barley stem gall midge undergoes two full generations and sometimes a third generation is completed if weather conditions are favorable. The first-generation adults emerge ~2 wk after the first significant rainfall, which normally corresponds to seeding time in late autumn. The second generation requires 2 mo to complete, and third instars of this generation may enter diapause. Adult flies are not known to feed and live only a few days. Mating may occur within 30–60 min after emergence from the puparia. Oviposition begins a few hours later and may extend over 2 d. Eggs are laid in the grooves between the longitudinal veins of leaves. Newly emerged larvae crawl down the leaves to the base of the stem, where they begin to feed. About 7% of the larvae fail to pupate. Diapausing third instars may survive more than a year and pupate only during the next favorable crop season.

Management. Planting barley after the peak adult emergence allows the crop to escape severe infestation. Adequate fertilization enhances tillering and plant vigor and increases plant tolerance to the pest. Chemical control is effective, but it is not normally economically feasible. Several sources of plant resistance have been identified in cultivated and wild barleys. These sources are being used in breeding programs to develop resistant barley cultivars.

Selected References. 58, 107, 109

Barley Shoot Fly

Scientific Classification. *Delia* spp. Diptera (Anthomyiidae). Different species of this genus have been reported in various parts of the world. *D. platura* (Meigen), formerly *Hylemia cilicrura* Rondani, is found in Morocco and Libya. *D. arambourgi* (Seguy) is reported in Ethiopia. Plant damage in Libya and Ethiopia is very similar suggesting that *D. platura* and *D. arambourgi* may be the same species. (Also see section on seedcorn maggot, pp. 67–68.)

Origin and Distribution. The barley shoot fly is widespread in Saharan Africa from Morocco to Ethiopia.

Description. Adult *D. platura* are grey-yellow and ~1/6–1/4 in. (4.2–6.4 mm) long. The compound eyes of the female are widely separated by a reddish band. The thorax is short, circular and grey-yellowish. The abdomen is narrow, grey and has 5 segments. The white eggs are ~1/25 in. (0.5 mm) long. The larvae are also white and measure ~1/4–1/3 in. (6.3–8.5 mm).

Pest Status. The barley shoot fly is widespread and damaging in Ethiopia and Libya, but occurs at sub-economic levels in the other North African countries.

Injury. The preferred hosts of the barley shoot fly are tef and barley, followed by wheat and oats. The young larvae bore down through the tissue to the growing point, killing the central shoot. One larva will destroy 3–4 shoots during its development. Yield losses up to 90% have been recorded on early-sown barley in Ethiopia and Libya.

Life History. *D. platura* lay up to 100 eggs in soil during a 2-wk oviposition period. The insect has 3 larval stages and pupates in the soil. Two to 5 generations may develop in a Mediterranean climate. *D. arambourgi* undergoes 2 generations on barley in Ethiopia.

Management. Several tolerant barley landraces have been identified in Ethiopia. Delaying planting until after peak fly emergence also reduces damage. Adequate fertilization stimulates tillering and compensates for fly-killed tillers. Imidacloprid applied as a seed dressing provides good control.

Selected References. 60, 81

Black Fly

Scientific Classification. *Phorbia securis* Tiensuu. Diptera (Anthomyiidae). Also known as the grey fly.

Origin and Distribution. The black fly is a pest of

(Top) Barley shoot fly adult (ICARDA). (Bottom) Barley damaged by the barley shoot fly (ICARDA).

wheat in Europe and North Africa. Other hosts include barley, oats, and *Bromus* spp.

Description. The adult is 1/6–1/5 in. (4.2–5.1 mm) long and is black. Antennae are 3-segmented; third segment is twice as long as the second. When newly laid, the egg is white and gradually yellows as it matures. The first instar larva is white and ~1/32 in. (0.8 mm) long. Larvae increase to 1/4–1/3 in. (6.4–8.5 mm) long as third instars.

Injury. Larvae feed inside the shoot, which becomes yellow and dies. The surrounding leaves are unaffected and remain green. In central Morocco, infestations as high as 34% have been reported and have forced farmers to replant. High infestation levels also have been observed in Algeria.

Life History. The black fly undergoes 1–2 generations per year. In France, only 1 generation develops between March and June; whereas in central Europe, 2 generations occur from September to October and from April to May. In Morocco, 2 generations develop from November to January and from February to May. The first generation is the

Pest Information

(Top) Black fly larva in wheat (ICARDA). (Bottom) Black fly larval damage to wheat (ICARDA).

most damaging because feeding occurs on young wheat plants. During her life span of 15–20 d, a female black fly can lay 2–3 eggs per day on the leaves. The incubation period takes 3–8 d depending on the temperature. The first instars enter the shoots and begin to feed. A larva generally completes its development in a single shoot. Most larvae pupate inside the stem at the base of the plant, although some pupate in the soil. Aestivation mostly occurs during the pupal stage of the second generation.

Management. Practical means of management are cultural methods that use crop rotation to avoid cereal monocultures and fertilization to favor tillering.

Selected References. 108

Migratory Locust

Scientific Classification. Desert locust, *Schistocerca gregaria* (Forskal) and African migratory locust, *Locusta migratoria migratoides* (Reiche and Fairmaire). Orthoptera (Acrididae).

Origin and Distribution. Locusts range across the African Sahel, which extends from the west coast of Africa south of the Sahara Desert to eastern Africa. This area is characterized by unpredictable seasonal rainfall and frequent, severe droughts. Locusts also are found on the Arabian and Indian Peninsulas. Areas affected by invading locust swarms include countries in northwest and eastern Africa, the Middle East, and southwest and south-central Asia.

The breeding range of migratory locusts is a very large region that encompasses the countries of Niger, Mali, Mauritania, Morocco, Algeria, Tunisia, Libya, Egypt, Sudan, and Saudi Arabia. This region also constitutes the "winter–spring" breeding area, where, if reproduction is successful, high populations will threaten countries of the Sahel and southwest Asia during the summer. Specific areas also are located in the winter–spring and summer reproduction areas of an invasion zone from which gregarious swarms depart en masse. These regions are the Occidental Region located in the extreme south of Algeria, south Mauritania, northern Mali, and northern Niger; the Central Region including the countries of eastern Africa and the Middle East; and the Oriental Region including India, Pakistan, Iran, and Afghanistan.

Description. Individuals of both species are large, brownish grasshoppers ranging from 2 to 3 in. (5–7.6 cm) long. Both species are polymorphic; there is a solitary and gregarious phase that may predominate in the population. Solitary-phase locusts are predominantly green; whereas those of the gregarious phase are primarily brown, yellow, or black-grey with long, strong wings as adults.

Pest Status. Locusts have been pests since ancient times. However, a period of relative absence occurred 1962 to 1986. Locust swarms reappeared in 1987 and 1989 and have since continued to be a threat throughout most of their range. The desert locus is the more serious pest of the two species.

Injury. Locusts are voracious plant feeders, devouring everything green and totally denuding a heavily infested area.

Life History. Locusts may produce up to 3 generations per year. Adults and nymphs may be captured year round. The first nymphs generally appear soon after winter or monsoon rains. They emerge either from eggs deposited in the ground the previous season by adults that died during the ensuing dry period, or from eggs laid at the beginning of the rainy period by adults that survived the dry period. The number of nymphal stages varies between 5 in males and 6 in females in the solitary, non-migratory phase. Males and females in the gregarious phase undergo 5 stages.

Both species prefer wet and moderately humid habitats with heavy vegetation. Development and dispersal of the gregarious phase are associated with the quality and quantity of vegetation and by the prevailing weather conditions, mainly temperature and wind. These conditions determine not only the presence or absence of flight activity, but also the distance flown per day. Migration will not start unless air temperature exceeds a critical threshold of 68–77 °F (20–25 °C). Transformation from the solitary phase to the gregarious phase gradually occurs in the population when locust density reaches ~12–24 per square yard (10–20/m^2).

Management. Because of the international scope of locust swarms, many regional and international organizations work together to control them. Vegetation development in locust breeding grounds is monitored by ground and satellite surveys, and swarms are tracked by radar. Meteorological stations throughout the locust's range monitor for weather conditions that are conducive to swarm formation and dispersal. The U.N. Food and Agriculture Organization (FAO) coordinates all desert locust control and works with other regional organizations and national governments that implement control measures. Insecticides are widely used for control and are applied as locust densities begin to rise.

Selected References. 52, 53, 54, 161

By Ross H. Miller, Mustapha El Bouhssini, and Saadia Lahloui

Beneficial Organisms

Entomopathogens for Control of Cereal Pests

Arthropod pests are affected by one or more entomopathogens, including bacteria, fungi, nematodes, protozoa, and viruses. Microbial control of a few cereal pests, such as chinch bug and plant bugs, has been attempted in crops. Yet, little work has been done on entomopathogens in cereals; and this work has focused on aphids, particularly the Russian wheat aphid.

Only fungi are known to cause disease outbreaks among aphids and are promising agents for aphid control. This is due to the ability of fungi to penetrate the aphid's cuticle. The piercing–sucking mouthparts of aphids preclude the primary, oral mode of entry of other pathogens. Few aphid pathogenic fungi pose any risk to vertebrates or plants, but some can infect nontarget insect species, such as ladybird beetles.

About 26 species of aphid pathogens occur in two groups of fungi, the Entomophthorales and the Hyphomycetes. Surveys in North America have found two hyphomycetes and eight entomophthoralean species infecting cereal aphids. Results in Eurasia and South Africa were similar.

Numerous fungal pathogens isolated from cereal aphids have been tested in the laboratory. Some Entomophthorales are difficult and costly to culture, limiting their use in pest management. The virulence of fungal species and that of isolates derived from a single species to a specific host vary greatly. Similarly, the virulence of a single fungal isolate varies among hosts. Careful screening of fungal strains against a target pest could enhance the potential for successful microbial control.

Low levels of infection and aphid mortality (enzootics) are common. Major disease outbreaks (epizootics) have been reported to kill up to 90% of a cereal aphid population, but these events are sporadic and often occur too late in the season to offer useful control. The fungi most commonly associated with epizootics in cereal aphid populations are *Pandora neoaphidis* (Remaudiere & Hennebert) Humber, *Conidiobolus obscurus* (Constantin) Batko, *Entomophthora chromaphidis* Burger & Swain, and *Neozygites fresenii* (Nowakowski) Batko. The relative importance of these fungi varies with aphid species.

Epizootics occur when high densities of aphids coincide with the presence of highly infectious pathogens under moderate temperatures and prolonged periods of high relative humidity or moisture. Epizootics typically occur during unusually wet weather in mid-to-late summer.

Aphid pathogenic Entomophthorales are more numerous and often more host specific than Hyphomycetes. Greater host specificity reduces the risk of impact on nontarget organisms including beneficial insects, such as ladybird beetles. Thus, host-specific strains are better candidates for use in classical biological control. Still, the

Russian wheat aphid 5 h after death due to *Beauveria bassiana* (Mingguang Feng).

Russian wheat aphid 48 h after death caused by *Beauveria bassiana* (Mingguang Feng).

Beneficial Organisms

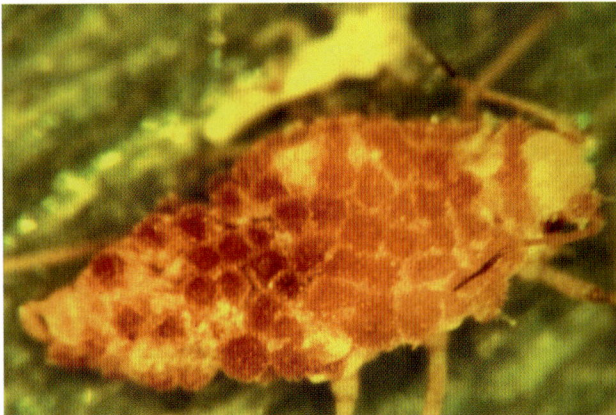

(Top and bottom) Russian wheat aphid 48 h after death caused by *Pandora neoaphidis* (Mingguang Feng).

Russian wheat aphid mummy (Keith S. Pike).

costs of preliminary safety tests exceed $100,000. To date, only one isolate each of *Beauveria bassiana* (Balsumo) Vuillemin, *Pandora neoaphidis*, *Paecilomyces fumosoroseus* (Wize) Brown & Smith, and *Zoophthora radicans* (Brefeld) Batko has been released experimentally in North America for control of the Russian wheat aphid.

Augmentative applications of several aphid-pathogenic fungi have been made. Inoculative and inundative treatments have been tested. Storage and application of many fungi are facilitated by their ability to desiccate and then resume development when rehydrated. The utility of fungi is also increased by the ability to apply hyphae for pest control. It is often quicker and less expensive to produce hyphae because not all fungi sporulate in culture, and spores of some species cannot be formulated for spray application.

Desiccated fungi can be applied as dusts or as granules. Alternatively, the spores and hyphae can be suspended in an aqueous solution and sprayed. More complex formulations, such as pellets, can protect the fungus from solar radiation and aid in rehydration. Host-specific fungi also could be applied to aphid parasitoids for delivery to the target pest.

As yet, fungal applications have not been highly successful for aphid control. Moderate rates of infection have been achieved shortly after application; however, induction of epizootics has been unreliable. It may be possible, in some cases, to enhance fungal activity, for example, by timing irrigation to increase moisture levels at the time of fungal application.

Fungus-based biological control of cereal aphids is not practical because of cost, reliability, and the time required to control the pests. It is possible that targeting oversummering or overwintering populations could be effective at reducing early-season crop infestations without increasing the risk of pesticide resistance or killing aphidophages. Further study of fungal ecology, formulations, and manipulations in the field is needed if fungal pathogens are to be exploited to control aphid pests.

Selected References. 55, 84, 101, 162

By James B. Johnson, Tadeusz J. Poprawski, and Stephen P. Wraight

Parasitoids (Hymenoptera) of Small Grain Insect Pests

Many small grain insect pests (e.g., aphids, armyworms, cereal leaf beetle, cutworms, grasshoppers, Hessian fly, wheat stem sawfly, white grubs) in North America are attacked by hymenopteran parasitoids. The life cycle and importance of most of these parasitoids have not been studied extensively. Some classical biological control has been attempted using introduced parasitoids with varying success on aphids, Hessian fly, and cereal leaf beetle. A synopsis of these insects and their parasitoids follows.

Grain Aphid Parasitoids

Scientific Classification. The primary parasitoids of small grain aphids in North America are a mixture of native and introduced species and belong to two families of Hymenoptera, the Aphelinidae (chalcidoid wasps, 3 species in the genus *Aphelinus*) and Aphidiidae (ichneumonoid wasps, 15 species in 6 genera: *Aphidius, Diaeretiella, Ephedrus, Lysiphlebus, Monoctonus,* and *Praon*).

Origin and Distribution. Most of the parasitoid species (9 of 15) are native to North America: *Aphelinus varipes* (Foerster), *Aphidius avenaphis* (Fitch), *Lysiphlebus testaceipes* (Cresson), *Ephedrus* sp. (prob. *californicus* Baker), *Monoctonus washingtonensis* Pike & Starý, *Praon americanum* Smith, *P. occidentale* Baker, *P. unicum* Smith, and *P. yakimanum* Pike & Starý. Six species are introduced, principally from Eurasia: *Aphelinus albipodus* Hayat & Fatima, *Aphelinus asychis* Walker, *Aphidius ervi* Haliday, *Aphidius colemani* Viereck, *Aphidius matricariae* Haliday, *Diaeretiella rapae* (M'Intosh). The introduced species have been recovered from various release areas in the United States; but only two of these species, *Aphidius ervi* and *D. rapae*, are commonly encountered. The aphelinids are found primarily in regions of semihumid and humid habitats. Of the native aphidiids, *A. avenaphis, L testaceipes, Praon occidentale,* and *P. unicum* are the most widely distributed across North America.

Description. Small wasps, adults range in size from <1/32 in. (<0.8 mm) (Aphelinidae) to 1/32–1/12 in. (0.8–2.1 mm) (Aphidiidae). Wasp coloration predominately is dark brown to black, with some more or less orange, yellow or yellow-brown aspects. In Aphidiidae, forewing anal vein present, antenna with more than 12 segments; Aphelinidae, forewing anal vein absent, antenna with 6 segments. Members of both families are solitary endophagous parasitoids.

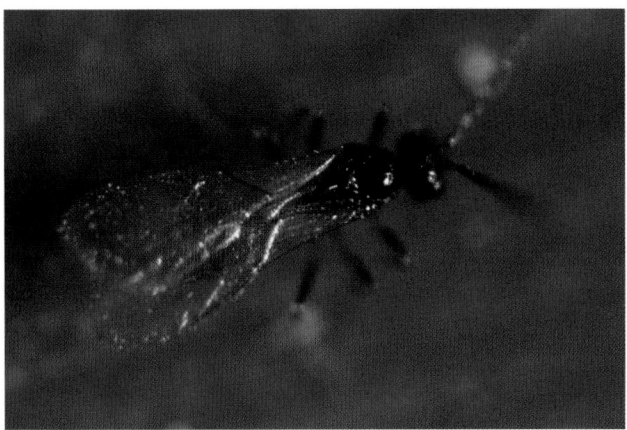

Adult aphid parasitoid, *Diaeretiella rapae* (Keith S. Pike).

Seasonal History and Life cycles. The occurrence of the parasitoid is generally synchronized with that of its aphid host species; interaction between parasitoid and aphid are common during much of the growing season. Environmental factors can disrupt the interactions; for example, at harvest when grain aphids are scarce, parasitoids are forced to move to an alternative aphid or enter seasonal diapause. Multiple generations occur annually, the number varies as a function principally of host availability and weather. Aphelinidae and Aphidiidae differ in several important biological ways, but especially in host feeding by adults. Aphidiids use their hosts only for oviposition. Aphelinids oviposit and feed on body fluids of the aphid; females puncture aphids with their ovipositors and then feed on the oozing haemolymph.

Parasitoid–Aphid Host Associations. All of the parasitoids of grain aphids are oligophagous; they attack two or more aphid hosts, but not equally. The parasitoid preferences (based on parasitoid–species population percentages, multiyear field findings, in the northwest United States) for grain aphids are (parasitoid–host): *A. avenaphis–S. avenae; A. ervi–S. avenae; A. albipodus–, A. asychis–,* and *A. varipes–D. noxia; D. rapae–D. noxia; L. testaceipes–R. maidis, –R. padi,* and *–S. graminum*. Parasitoid species not mentioned are of minor importance (few and/or infrequent in occurrence).

Parasitoid Effects on Aphids and Aphid Populations. The primary effects of parasitization on the aphid are (1) punctured cuticle (may be lethal), (2) starvation and death due to internal larval feeding on haemolymph and tissues (3) reduced fecundity generally, (4) increased food consumption, reduced assimilation, (5) increased honeydew output, (6) slowed development rate (particularly if aphids are parasitized in early instars), (7) alarm pheromones sometimes released, often causing dispersal, and (8) changed aphid color, shape and size compared with unparasitized aphids. Generally parasitoid action does not eradicate aphid populations, but the reduction in numbers of aphids can be dramatic. The performance or action levels of parasitoids can be reduced by aphid predators, secondary parasitoids, and fungal diseases, and by adverse weather.

Cereal Leaf Beetle Parasitoids

The cereal leaf beetle has come under excellent biological control in the north-central and Great Lakes states, and Utah by virtue of a complex of parasitic wasps [*Anaphes flavipes* Forster, *Diaparsis* spp., *Lemophagus curtus* Townes, and *Tetrastichus julis* (Walker)] introduced from Europe.

Beneficial Organisms

Tetrastichus julis, parasitoid of cereal leaf beetle larva (Mark Hitchcox).

Populations of the beetle, which are constantly low now, are no longer managed by pesticides. *T. julis* is perhaps the keystone species of the complex; it shows common seasonal presence and population stability, and usually, but not always, exerts the highest levels of control compared with the other species. Currently, biological control is inadequate in Virginia and North Carolina, and in parts of the western United States (Idaho, Montana, and Wyoming).

Hessian Fly Parasitoids

Thirty-five hymenopterous parasitoids attack Hessian fly in North America—only several are important. One exotic species, *Pediobius* (=*Pleurotropis*) *metallicus*, was introduced into the United States from England in the 1890's, and is of minor importance in the eastern and central United States. The only parasitoid that attacks Hessian fly in autumn is *Platygaster hiemalis* Packard. This parasitoid is polyembryonic and produces multiple parasitoids from each host puparium. It mainly occurs in the eastern United States, but recently it was introduced into Texas from Indiana. All other species attack Hessian fly during the spring. *Homoporus destructor* (Say) and *Eupelmus allynii* (French) typically are the most prevalent spring parasitoids throughout North America. *Eupelmella vesicularis* (Retzius), *Trichomalopsis subapterus* (Riley), *T. americanus* Gahan, *Pediobius epigonus* Walker, *Platygaster herrickii* Packard, *P. vernalis* Myers, and *Pediobius metallicus* (Nees) may be prevalent in certain areas. All spring parasitoids are solitary parasites of Hessian fly puparia, except *P. vernalis*, which is polyembryonic. As a group, the parasitoids may kill routinely up to 70–80% of Hessian fly during the spring. Parasitism of Hessian fly in autumn by *P. hiemalis* varies but can exceed 50% in some years. Parasitoids are believed to play an important role in the long-term regulation of Hessian fly populations, but they probably have minimal effect on the numbers of Hessian fly within a growing season.

Selected References. 32, 158, 183

By Keith S. Pike, Petr Starý, G. David Buntin, and Jay B. Karren

Predators

Small grain fields harbor many species of predatory arthropods. The role of most predatory species in controlling insect pests of small grains in North America is not well understood and more research is needed.

Lady Beetles

Species in the genera *Adalia, Coccinella, Coleomegilla, Hippodamia, Olla,* and *Scymnus*. Coleoptera (Coccinellidae).

Lady beetles occur in small grains throughout North America. Most species are native to North America including the twelvespotted lady beetle, *Coleomegilla maculata* (DeGeer), and convergent lady beetle *Hippodamia convergens* Guérin-Méneville. Some exotic species such as the sevenspotted lady beetle, *Coccinella septempunctata* L., and multicolored Asian lady beetle, *Harmonia axyridis* (Pallas), have been established. Other less abundant species include

(Left) Convergent lady beetles, *Hippodamia convergens* (D. A. Beck). (Right) Lady beetle, *Hippodamia sinuata* (R. W. Behle).

Beneficial Organisms

Lady beetle larva and aphid prey (R. D. Eikenbary).

Adalia bipunctata (L.), *Cycloneda munda* (Say), *C. polita* (Casey), *Hippodamia sinuata* Mulsant, *H. parenthesis* (Say), *Olla v-nigrum* (Muslant), several *Scymnus* spp., and others. Coccinellids prey primarily on aphids, but may also attack spider mites, other insect larvae and insect and mite eggs. Several *Stethorus* spp. prey on spider mites and sometimes inhabit small grain fields.

Because of the wide climatic and geographic conditions under which small grains are grown, and the complexity of predator–prey relationships, general statements about coccinellid effects on aphid infestations are difficult to make. Under favorable conditions coccinellids are voracious predators and can drive an aphid infestation to extinction. Late instars of *H. convergens* can consume up to 40 greenbugs per day. Prey density is key to retaining adult coccinellids in a small grain field. In order not to oviposit in a low density or declining aphid infestation, thereby stranding larvae with insufficient food, adult lady beetles will disperse to search for better sources of prey. Adults must consume a certain amount of aphid biomass before egg production begins, therefore low aphid densities in an area will increase the coccinellids' mean generation time. Fortunately, most coccinellids prey on several aphid species. If more than one aphid species occur concurrently in a small grain field, they act for all practical purposes as a single food source for coccinellids.

Flower Flies and Aphid Flies

Several species of flower flies, Diptera (Syrphidae). Several species of aphid flies, Diptera (Chamaemyiidae).

Flower fly adults, often called hover flies or syrphid flies, typically are black with yellow bands and markings that mimic wasps or bees. Aphid flies are small and gray with black spots on the abdomen. Adults are not predatory and feed on nectar, pollen, and aphid honeydew. Larvae

Adult flower fly (V. H. Beregovoy).

of flower fly and aphid fly are maggots that actively crawl over cereal plants and prey on aphids. They are generally greenish, sometimes with pale white stripes. Larvae are usually predaceous on aphids and can be common when aphids are present.

The importance of flower and aphid flies in biological control of cereal aphids in the United States has not been determined. Studies in Europe indicate that flower fly larvae are important in suppressing cereal aphid populations. Providing flowering plants in the field margin can enhance reproduction by flower flies in small grain fields.

Lacewings

Common green lacewing, *Chrysoperla plorabunda* (Fitch) and *C. carnea* (Stephens), Neuroptera (Chrysopidae); *Hemerobius pacificus* (Banks), Neuroptera (Hemerobiidae).

Only a few lacewing species inhabit small grain fields in North America. Green lacewings are common and widespread species. *Hemerobius pacificus* is often called the

Lacewing larva (V. H. Beregovoy).

Beneficial Organisms

brown lacewing and occurs in western North America. Adult lacewings are highly mobile and reproduce preferentially in habitats with abundant prey. Adults are ~1 in. (2.5 cm) long, generally green or yellow whereas brown lacewings are smaller and brown. The wings are lacy and fold roof-like over the body. Adults feed on aphid honeydew, nectar, and pollen. Green lacewing eggs generally are white and placed on top of a slender silk stalk. Brown lacewings lay eggs directly on the plant.

Lacewings that inhabit small grains are predaceous only as larvae. Lacewing larvae are elongate, flat and sicklelike mandibles project forward from the head. Larvae feed by piercing prey to suck body fluids of small, soft-bodied arthropods, including aphids, plant bugs, thrips, mites, and the eggs and larvae of Lepidoptera and Diptera. Some nonagricultural habitats and fields of agricultural crops serve as reservoirs from which lacewings colonize small grains. The abundance and diversity of such habitats and their distance from small grain fields are important factors influencing the density of lacewings.

Minute Pirate Bugs and Flower Bugs

Several species in the genus *Orius*, Hemiptera–Heteroptera (Anthocoridae).

Two species of *Orius* inhabit small grains, *O. insidious* (Say) in the eastern and *O. tristicolor* (White) in the western North America. *O. minutus* (L.) is a predator of grain insects in Europe. It was introduced accidentally and became established in the Pacific Northwest. Its range is expanding slowly eastward.

Adult *Orius* are 1/8 in. (3.2 mm) long and appear black and white. Adults overwinter in leaf litter. Eggs are laid in plant tissue. Nymphs are oval and yellow-orange but become dark brown as fifth instars. *Orius* mainly feed on thrips and aphids, but also may feed on moth eggs and small caterpillars. Their small size allows them to reach prey in curled leaves and elsewhere on plants that are not accessible to larger predators. Several other species of *Orius* are well known in small grain fields in Europe and are likely candidates for biological control in North America. At least one is reported to feed on Russian wheat aphid.

Other Predatory Bugs

Damsel bugs, *Nabis* spp. (Nabidae); stink bugs, *Podisus maculiventris* (Say) and others (Penatomidae); and various species of assassin bugs (Reduviidae); Hemiptera–Heteroptera.

Adult damsel bugs are typically tan, brown or black

Damsel bug adult (V. H. Beregovoy).

and 1/5–1/2 in. (5.1–12.7 mm) long. Assassin bugs are variable in size usually >1/2 in. (12.7 mm) long and brown or black. Predatory stink bugs usually are brown and resemble plant-feeding stink bugs except the beak is 3–4 times wider than in plant-feeding types.

Adults of damsel, assassin, and predatory stink bugs are winged and highly mobile. Damsel bugs feed on small, soft-bodied arthropods in small grain fields. Assassin bugs and predatory stink bugs typically feed on larger insects including caterpillars and other slow-moving prey. Adults and immatures are predaceous. The damsel bug, *Nabis americoferus* Carayon, is widely distributed in small grains and other crops and is frequently among the most abundant predatory insects in small grain fields.

Damsel bugs and other predatory bugs are opportunistic, general predators and not considered effective at controlling any particular pest because they lack the prey specificity. However, they may be important because of their high abundance in small grains. *N. americoferus*, and perhaps other predatory bugs, respond positively to habitat diversification within fields and in the surrounding landscape.

Ground and Rove Beetles

Many species of ground beetles, Coleoptera (Carabidae); many species of rove beetles, Coleoptera (Staphylinidae).

Ground beetles range from ~1/4 in. (6.2 mm) to >1 in. (2.5 cm) long. They usually are flattened somewhat and brown or black. Most species have distinct mandibles projecting forward from the head. Rove beetles are smaller ~1/2 in. (12.7 mm) or less, black or brown with slender or tear drop-shaped bodies. Rove beetles have very short wing covers; most of the abdomen is exposed.

Beneficial Organisms

A ground beetle (V. H. Beregovoy).

Spiders: Line-weaving spider web (top, M. Greenstone) and male (middle, M. Greenstone) and greenbug in web of *Tetragnatha laboriosa* (bottom, Scott Bauer).

Ground and rove beetles are abundant predators in small grain fields. They feed primarily at the soil surface and belowground and hence occupy a different feeding niche than other insect predators, such as lady beetles that forage primarily on plants. Ground and rove beetles prey on a broad range of invertebrates inhabiting the soil or on the soil surface, although some species are omnivores or plant seed feeders. More than 60 species of ground and rove beetles can be found in a single small grain field over the course of the growing season. Species differ widely in size, abundance, and seasonal activity pattern. As a group, they are capable of exploiting a broad range of prey of varying size and seasonal occurrence.

The abundance and species diversity of ground and rove beetles in small grain fields are related to within-field and extra-field management practices. Crop production practices such as cultivation method and crop rotation can influence populations of these predators. In Europe, providing grassy strips in small grain fields and field margins as overwintering habitat increases the abundance of ground and rove beetle in small grain fields during the growing season.

Spiders

Many species of spiders, Araneae (several families).

Spiders have not been studied extensively in small grains in the United States. An intensive study in southeastern Colorado indicates that they are much less abundant there than in Great Britain, and more evenly represented by different families. For example, although the Linyphiidae (line-weaving spiders) dominate in both places, they represented only 22% of individuals in Colorado, compared with 71–97% at one intensively studied site in Great Britain. Lycosidae (wolf spiders) made up 17% of Colorado spiders, Thomisidae (crab spiders) 12%, and Araneidae (typical orb weavers), Tetragnathidae (long-jawed orb weavers), Theridiidae (cobweb weavers) and Gnaphosidae collectively 9–10% of individuals. The vast majority feed on or near the soil surface. Most of the line-weaving spiders are minute, not more than 1/10 in. (2.5 mm) long and difficult to see; but their postage stamp-sized horizontal sheet webs are often very abundant and visible just above the soil surface. By killing large numbers of cereal aphids that fall from plants, they are important in reducing aphid populations in Europe; their importance in North America remains to be determined.

Degree of Natural Control

Although the importance of predators in insect pest management in small grains in North America has not been sufficiently studied, some research has focused on the effect of predatory arthropods on cereal aphid infestations. Field studies have shown that the complex of predators described above, often in conjunction with insect parasitoids, can control cereal aphid populations under some circumstances. Lady beetles are believed to be the most important cereal aphid predators in this complex, but more research is needed to delimit the roles of various predators in cereal aphid biological control.

Effects of Management

In pest management of cereal aphid, conservation of naturally occurring predators seems to be the best approach for incorporating them in an overall program. Lady beetles, lacewings, and most other aphid predators are very susceptible to conventional insecticide, especially organophosphates and pyrethroids. Therefore, chemical controls should be avoided unless predators are obviously incapable of suppressing an aphid infestation. Inundative releases of these natural enemies to control aphid infestations in small grains are not recommended because of high cost and the strong tendency of many predator species to disperse from fields in search of prey. The effects of cultural practices, such as tillage and planting time, on predator populations have not been studied extensively in small grains in North America.

Selected References. 29, 31, 63, 65, 79, 82, 102, 103, 104, 170, 185, 186, 188

By Gerald J. Michels, Jr., Matthew H. Greenstone, B. Wade French, Norman C. Elliott, and John D. Lattin

References

1. **Agricultural Statistics. 2005.** United States Department of Agriculture-National Agricultural Statistics Service (http://www.usda.gov/nass/pubs/agstats.htm).
2. **Aldrich, J. M. 1918.** Seasonal climatic variation in *Cerodontha*. Ann. Entomol. Soc. Am. 11: 63–66.
3. **Aldrich, J. M. 1920.** European frit fly in North America. J. Agric. Res. 18: 451–474.
4. **Allen, M. W. 1937.** Observations on the biology of the wheat-stem maggot in Kansas. J. Agric. Res. 55: 215–238.
5. **Allsopp, P. G. 1980.** The biology of false wireworms and their adults (soil inhabiting Tenebrionidae: Coleoptera): A review. Bull. Entomol. Res. 7: 343–379.
6. **Appleby, A. P. 1987.** Weed control in wheat, pp. 396–415. *In* E. G. Heyne et al. [Eds.]. Wheat and wheat improvement, 2nd ed. Agronomy Monograph 13, American Society of Agronomy, Madison, WI.
7. **Armitage, H. M. 1954.** Current insect notes. Bulletin of the California State Department of Agriculture. 43: 73–76.
8. **Bates, B. A., M. J. Weiss, R. B. Carlson, and D. K. McBride. 1991.** Sequential sampling plan for *Limothrips denticornis* (Thysanoptera: Thripidae) on spring barley. J. Econ. Entomol. 84: 1630 1634.
9. **Battenfield, S. L., S. G. Wellso, and D. L. Haynes. 1982.** Bibliography of cereal leaf beetle, *Oulema melanopus* (L.) (Coleoptera: Chrysomelidae). Bull. Entomol. Soc. Am. 28: 291–301.
10. **Benbrook, C. M. 1996.** Pest management at the crossroads. Consumer Union, Yonkers, NY.
11. **Beregovoy, V., and D. C. Peters. 1997.** Efficiency of host plant use by eight biotypes of *Schizaphis graminum* (Aphididae: Homoptera). J. Kan. Entomol. Soc. 69: 69–77.
12. **Binns, M. R., J. P. Nyrop, and W. van der Werf. 2000.** Sampling and monitoring in crop protection. CABI Publishing, New York.
13. **Blackman, R. L., and V. F. Eastop. 1984.** Aphids on the world's crops: an identification and information guide. John Wiley & Sons, New York.
14. **Blodgett, S. L., P. M. Denke, M. A. Ivie, C. W. O'Brien, and A. W. Lenssen. 1997.** *Listronotus montanus* Dietz (Coleoptera: Curculionidae) damaging spring wheat in Montana. Can. Entomol. 129: 377–378.
15. **Boeve, P. J., and M. J. Weiss. 1997.** Binomial sequential sampling plans for cereal aphids (Homoptera: Aphididae) in spring wheat. J. Econ. Entomol. 90: 967–975.
16. **Bonnemaison, L. 1980.** Principal animal pests, pp. 59-68. *In* E. Häfliger, [Ed.]. Wheat. Doucumenta CIBA–GEIGY, Basel, Switzerland.
17. **Brakke, M. K. 1987.** Virus diseases of wheat. pp. 585–603. *In* E. G. Heyne [Ed.]. Wheat and wheat improvement, 2nd ed. Agronomy Monograph 13, American Society of Agronomy. Madison, WI.
18. **Brewer, M. J., and N. C. Elliott. 2004.** Biological control of cereal aphids and mediating effects of host plant and habitat manipulations. Annu. Rev. Entomol. 49: 219–242.
19. **Briggle, L. W., and B. C. Curtis. 1987.** Wheat worldwide, pp. 1–32. *In* E. G. Heyne [Ed.]. Wheat and wheat improvement, 2nd ed. Agronomy Monograph 13. American Society of Agronomy, Madison, WI.
20. **Buntin, G. D. 1992.** Damage by the European corn borer (Lepidoptera: Pyralidae) to winter wheat. J. Entomol. Sci. 27: 361–365.
21. **Buntin, G. D., and R. J. Beshear. 1995.** Seasonal abundance of thrips (Thysanoptera) on winter small grains in Georgia. Environ. Entomol. 24: 1216 1223.
22. **Buntin, G. D., and J. W. Chapin. 1990.** Biology of Hessian fly (Diptera: Cecidomyiidae) in the southeastern United States: Geographic variation and temperature-dependent phenology. J. Econ. Entomol. 83: 1015–1024.
23. **Buntin, G. D., and J. K. Greene. 2004.** Abundance and species composition of stink bugs (Heteroptera: Pentatomidae) in Georgia winter wheat. J. Entomol. Sci. 39: 287–290.
24. **Buntin, G. D., K. L. Flanders, R. W. Slaughter, and Z. D. DeLamar. 2004.** Damage loss assessment and control of the cereal leaf beetle (Coleoptera: Chrysomelidae) in winter wheat. J. Econ. Entomol. 97: 374–382.
25. **Burn, A. J. 1987.** Cereal crops, pp. 209–256. *In* A. J. Burn, T. H. Coaker, and P. C. Jepson [Eds.]. Integrated pest management. Academic Press, London.
26. **Burton, R. L., K. J. Starks, and D. C. Peters. 1980.** The army cutworm. Okla. Agric. Exp. Stn. Bull. B-749.
27. **Calkins, C. O., and V. M. Kirk. 1973.** Distribution and movement of adult false wireworms in a wheat field. Ann. Entomol. Soc. Am. 66: 527–532.
28. **Calkins, C. O., and V. M. Kirk. 1975.** False wireworms of economic importance in South Dakota (Coleoptera: Tenebrionidae). USDA Research Service B633, Washington, DC.
29. **Canard, M., Y. Semeric, and T. R. New. 1984.** Biology of Chrysopidae. Dr. W. Junk, The Hague, The Netherlands.
30. **Capinera, J. L., and T. S. Sechrist. 1982.** Grasshoppers (Acrididae) of Colorado. Colo. State Univ. Exp. Stn. Bull. 584S.
31. **Carcamo, H. A., J. K. Niemala, and J. R. Spence. 1995.** Farming and ground beetles: effects of agronomic practice on populations and community structure. Can. Entomol. 127: 123–140.
32. **Carter, N., I. F. G. McLean, A. D. Watt, and A. F. G. Dixon. 1980.** Cereal aphids: a case study and review, pp. 271–348. *In* T. H. Coaker [Ed.]. Applied biology, vol. 5. Academic Press, New York.
33. **Carver, B. F., and James D. Ownby. 1995.** Acid soil tolerance in wheat, pp. 117–173. *In* Donald L. Sparks [Ed.]. Advances in agronomy, vol. 54. Academic Press, San Diego.
34. **Cate, P. 1980.** Corn ground beetle (*Zabrus tenebrioides* Goeze) an important pest of grain in Austria [in German]. Pflanzenschutz. 33: 115–117.
35. **Chamberlin, T. R. 1941.** The wheat jointworm in Oregon, with special reference to its dispersion, injury, and parasitization. U.S. Dep. Agric. Tech. Bull. 784.
36. **Chapin, J. W., J. S. Thomas, S. M. Gray, D. W. Smith, and S. B. Halbert. 2001.** Seasonal abundance of aphids in wheat and their role as barley yellow dwarf virus vectors in the South Carolina coastal plain. J. Econ. Entomol. 94: 410–421.
37. **Coffman, F. A. [ed.]. 1961.** Oats and Oat Improvement. Agronomy Monograph No. 6, American Society of Agronomy, Madison, WI

References

38. Collins, K. L., N. D. Boatman, A. Wilcox, J. M. Holland, and K. Chaney. 2002. Influence of beetle banks on cereal aphid predation in winter wheat. Agric. Ecosyst. Environ. 93: 337–350.
39. Cook, R. J., and R. J. Veseth. 1991. Wheat health management. APS (American Phytopathological Society) Press, St. Paul, MN.
40. Cox, T. S. 1991. The contribution of introduced germplasm to the development of U.S. wheat cultivars, pp. 25–48. In H. L. Shands and L. E. Wiesner [Eds.]. Use of plant introductions in cultivar development Part 1. Crop Science Society of America, Madison, WI.
41. Crocker, R. L., D. Marshall, and J. S. Kubica-Breier. 1990. Oat, wheat, and barley resistance to white grubs of *Phyllophaga congrua* (Coleoptera: Scarabaeidae). J. Econ. Entomol. 83: 1558-1562.
42. Crumb, S. E. 1929. Tobacco cutworms. U.S. Dep. Agric. Tech. Bull. 88.
43. D'Arcy, C. J., and P. A. Burnett, [Eds.]. 1995. Barley yellow dwarf: 40 years of progress. APS (American Phytopathological Society) Press, St. Paul, MN.
44. Davidson, R. H., and W. F. Lyon. 1987. Corn leaf aphid, p. 194. In Insect pests of farm, garden and orchard, 8th ed. John Wiley & Sons, New York.
45. Delong, D. M. 1948. The leafhoppers, or Cicadellidae, of Illinois. Ill. Nat. Hist. Surv. Bull. 24, Article 2.
46. Denys, C., and T. Tscharntke. 2002. Plant–insect communities and predator–prey ratios in field margin strips, adjacent crop fields, and fallows. Oecologia 130: 315–324.
47. Doane, J. F., O. Olfert, and M. K. Mukerji. 1987. Extraction precision of sieving and brine flotation for removal of wheat midge, *Sitodiplosis mosellana* (Diptera: Cecidomyiidae), cocoons and larvae from soil. J. Econ. Entomol. 80: 268–271.
48. Duran, M. 1971. Investigations on ground pearls [*Margarodes* (*Porphyrophora*) *tritici* Bodenheimer], a grain pest in central Anatolia [in Turkish with English summary]. Bitki Koruma Bulteni V.1, Suppl. 1. Yeni Desen Matbaasi.
49. Elliott, R. H. 1988. Factors influencing the efficacy and economic returns of aerial sprays against the wheat midge, *Sitodiplosis mosellana* (Géhin) (Diptera: Cecidomyiidae). Can. Entomol. 120: 941–954.
50. Elliott, R. H., and L. W. Mann. 1996. Susceptibility of red spring wheat, *Triticum aestivum* L. cv. Katepwa, during heading and anthesis to damage by wheat midge, *Sitodiplosis mosellana* (Géhin) (Diptera: Cecidomyiidae). Can. Entomol. 128: 367–375.
51. Elliott, R. H., and L. W. Mann. 1997. Control of wheat midge, *Sitodiplosis mosellana* (Géhin), at lower chemical rates with small-capacity sprayer nozzles. Crop Prot. 16: 235–242.
52. FAO (Food and Agriculture Organization of the United Nations). 1962. Rapport d'une reunion sur les problemes meteorologiques interessant le projet relatif au criquet pelerin, October 1962. Rome.
53. FAO (Food and Agriculture Organization of the United Nations). 1992. Rapport de la commission de lutte contre le criquet pelerin en Afrique du Nord-Ouest. 18th session, 24–29 Oct. 1992. Algers, Algeria.
54. FAO (Food and Agriculture Organization of the United Nations). 1995. Rapport de la quatrieme session du groupe technique sur le criquet pelerin, March 21–24, 1995. Rome.
55. Feng, M-G., R. M. Nowierski, J. B. Johnson, and T. J. Poprawski. 1992. Epizootics caused by entomophthoralean fungi (Zygomycetes, Entomophthorales) in populations of cereal aphids (Hom., Aphididae) in irrigated small grains of southwestern Idaho, USA. J. Appl. Entomol. 113: 376–390.
56. Fenster, C. R., C. E. Domingo, and O. C. Burnside. 1969. Weed control and plant residue maintenance with various tillage treatments in a winter wheat–fallow rotation. Agron. J. 61: 256–259.
57. Funderburk, J. E., D. C. Herzog, R. K. Sprenkel, and R. E. Lynch. 1984. Parasitoids and pathogens of larval lesser cornstalk borer (Lepidoptera: Pyralidae) in northern Florida. Environ. Entomol. 13: 1319–1323.
58. Gagné, R. J., J. H. Hatchett, S. Lhaloui, and M. El Bouhssini. 1991. Hessian fly and barley stem gall midge, two different species of *Mayetiola* (Diptera: Cecidomyiidae) in Morocco. Ann. Entomol. Soc. Am. 84: 436–443.
59. Ganassi, S., A. Moretti, A. M. B. Pagliai, A. Logrieco, and M. A. Sabatini. 2002. Effects of beauvericin on *Schizaphis graminum* (Aphididae). J. Invert. Path. 80: 90–96.
60. Gebre, H., and J. A. G. van Leur [Eds.]. 1996. Barley research in Ethiopia: Past work and future prospects, pp. 128–137. In Proceedings of the first barley research review workshop, 16–19 October, Addis Ababa: IAR/ICARDA. Addis Ababa, Ethiopia.
61. Gilbertson, G. I. 1925. The wheat-stem maggot. S.D. Agric. Exp. Stn. Bull. 217.
62. Giles, K. L., D. B. Jones, T. A. Royer, N. C. Elliott, and S. D. Kindler. 2003. Development of a sampling plan in winter wheat that estimates cereal aphid parasitism levels and predicts population suppression. J. Econ. Entomol. 96: 975–982.
63. Gordon, R. D. 1985. The Coccinellidae (Coleoptera) of America north of Mexico. J. N.Y. Entomol. Soc. 93: 1–912.
64. Gray, S. M., G. C. Bergstrom, R. Vaughan, D. M. Smith, and D. W. Kalb. 1996. Insecticidal control of cereal aphids and its impact on the epidemiology of the barley yellow dwarf luteoviruses. Crop Prot. 15: 687–697.
65. Greenstone, M. H. 2001. Spiders in wheat: First quantitative data for North America. BioControl 46: 439–454.
66. Guppy, J. C. 1961. Life history and behavior of the armyworm, *Pseudaletia unipuncta* (Haw.) (Lepidoptera: Noctuidae), in eastern Ontario. Can. Entomol. 93: 1141–1153.
67. Gurr, G. M., S. D. Wratten, and P. Barbosa. 2000. Success in conservation biological control of arthropods, pp. 105–132. In G. Gurr, and S. Wratten [Eds.]. Biological control: measures of success. Kluwer Academic, London.
68. Halbert, S. E., and D. J. Voegtlin. 1995. Biology and taxonomy of vectors of barley yellow dwarf viruses, pp. 217–258. In C. J. D'Arcy and P. A. Burnett [Eds.]. Barley yellow dwarf: 40 years of progress. APS (American Phytopathological Society) Press, St. Paul, MN.
69. Hammond, R. B., and R. L. Cooper. 1993. Interaction of planting times following the incorporation of a living, green cover crop and control measures on seedcorn maggot populations in soybean. Crop Prot. 12: 539–543.
70. Harvey, T. L., T. J. Martin, and D. L. Seifers. 1994. Importance of plant resistance to insect and mite vectors in controlling virus diseases of plants: resistance to the wheat curl mite (Acari: Eriophyidae). J. Agric. Entomol. 1: 271–277.
71. Hatchett, J. H., and R. L. Gallun. 1970. Genetics of the ability of the Hessian fly, *Mayetiola destructor* (Say), to survive on wheat having different genes for resistance. Ann. Entomol. Soc. Am. 63: 1400–1407.
72. Hatchett, J. H., K. J. Starks, and J. A. Webster. 1987. Insect and mite pests of wheat, pp. 625–675. In E. G. Heyne et al. [Eds.]. Wheat and wheat improvement, 2nd ed. Agronomy Monograph 13, American Society of Agronomy, Madison, WI.
73. Haynes, D. L., and S. H. Cage. 1981. The cereal leaf beetle in North America. Annu. Rev. Entomol. 26: 259–287.
74. Heer, W. F., and E. G. Krenzer, Jr. 1989. Soil water availability for spring growth of winter wheat (*Triticum aestivum* L.) as influenced by early growth and tillage. Soil Tillage Res. 14: 185–196.
75. Henry, T. J., and J. D. Lattin. 1987. The taxonomic status, biological attributes, and recommendations for future work on the genus *Lygus* (Heteroptera: Miridae), pp. 54–68. In R. Hedlund and H. M. Graham [Eds.]. Economic importance and biological con-

References

trol of *Lygus* and *Adelphocoris* in North America. U.S. Dep. Agric. Publ. ARS 64.

76. **Herbert, D. A., Jr., and J. W. van Duyn.** 1999. Cereal leaf beetle: biology and management. Virginia Cooperative Extension System Publ. 444-350.

77. **Herrman, T. J., R. L. Bowden, T. Loughin, and R. K. Bequette.** 1996. Quality response to the control of leaf rust in Karl hard red winter wheat. Cereal Chem. 73: 235–238.

78. **Heyne, E. G. [Ed.].** 1987. Wheat and Wheat Improvement, 2nd ed. Agronomy Monograph 13, American Society of Agronomy, Madison, WI.

79. **Hickman, J. M., and S. D. Wratten.** 1996. Use of *Phacelia tanacetifolia* strips to enhance biological control of aphids by hoverfly larvae in cereal fields. J. Econ. Entomol. 89: 832–840.

80. **Higley, L. G., and L. P. Pedigo [Eds.].** 1996. Economic thresholds for integrated pest management. University of Nebraska Press, Lincoln.

81. **Hill, D. S.** 1983. Agricultural insect pests of the tropics and their control, 2nd ed. Cambridge University Press, Cambridge, UK.

82. **Hodek, I., and A. Honek.** 1996. Ecology of Coccinellidae. Kluwer Academic, Dordrecht, The Netherlands.

83. **Hoelscher, C. E., J. G. Thomas, and G. L. Teetes.** 1997. Aphids on Texas small grains and sorghum. Tex. Agric. Ext. Serv. Bull. B-1572. http://entowww.tamu.edu/extension/bulletins/b-1572.html

84. **Humber, R. A.** 1991. Fungal pathogens of aphids, pp. 45–56. *In* D. C. Peters, J. A Webster and C. S. Chouber [Ed.]. Aphid–plant interactions—populations to molecules. Proceedings of the OSU Centennial Event, Stillwater. Okla. St. Univ. Agric. Res. Stn. Rep. MP-132.

85. **Hunt, T. N., and J. R. Baker.** 1982. Insects and related pests of field crops. N.C. Agric. Ext. Serv. AG-271.

86. **Ihrig, R. A., D. A. Herbert, Jr., J. W. van Duyn, and J. R. Bradley.** 2001. Relationship between cereal leaf beetle (Coleoptera: Chrysomelidae) egg and fourth-instar populations and impact of fourth-instar defoliation of winter wheat yields in North Carolina and Virginia. J. Econ. Entomol. 94: 634–639.

87. **Jackson, K. E., J. P. Damicone, H. A. Melouk, P. W. Pratt, J. R. Sholar, P. Mulder, and B. Stacy.** 1995. Results of 1994 plant disease control field studies. Okla. Agric. Exp. Stn. Ext. Serv. Report 3321.

88. **Jacobson, L. A.** 1971. The pale western cutworm, *Agrotis orthogonia* Morrison (Lepidoptera: Noctuidae): a review of research. Quaest. Entomol. 7: 414–436.

89. **Jeppson, L. R., H. H. Keifer, and E. W. Baker.** 1975. Mites injurious to economic plants. University of California Press, Berkeley.

90. **Johnston, R. L., and G. W. Bishop.** 1987. Economic injury levels and economic thresholds for cereal aphids (Homoptera: Aphididae) on spring-planted wheat. J. Econ. Entomol. 80: 478–482.

91. **Kamm, J. A.** 1975. Sawflies in fine fescue grown for seed. Environ. Entomol. 4: 312–314.

92. **Kieckhefer, R. W.** 1974. Seasonal appearance and movement of adult stem maggots in South Dakota cereal crops. J. Econ. Entomol. 67: 558–561.

93. **Kieckhefer, R. W., and W. L. Morrill.** 1970. Estimates of loss of yield caused by the wheat stem maggot to South Dakota cereal crops. J. Econ. Entomol. 63: 1426–1429.

94. **Knowlton, G. F.** 1931. The wheat strawworm, *Harmolita grandis* Riley, in Utah—1930. J. Econ. Entomol. 24: 414–416.

95. **Knowlton, G. F., and M. J. Janes.** 1933. Distribution and damage by jointworm flies in Utah. Utah Agric. Exp. Stn. Bull. 243.

96. **Knowlton, G. F., and F. V. Lieberman.** 1954. Controlling the wheat strawworm. Utah St. Agric. Ext. Serv. Circ. 194.

97. **Knutson, A. E., E. A. Rojas, D. Marshal, and F. E. Gilstrap.** 2002. Interaction of parasitoids and resistant cultivars of wheat on Hessian fly, *Mayetiola destructor* Say (Cecidomyiidae). Southwest. Entomol. 27: 1–10.

98. **Krazheva, L. P., E. A. Pnomareva, and U. N. Chikhacheva.** 1981. Protection of wheat from the carabid beetle *Zabrus tenebrioides*, USSR. Zash. Rast. (Moscow) 7: 44–45.

99. **Landis, D. A., S. D. Wratten, and G. M. Gurr.** 2000. Habitat management to conserve natural enemies of arthropod pests in agriculture. Annu. Rev. Entomol. 45: 175–201.

100. **Langhof, M., A. Gathmann, H. M. Poehling, and R. Meyhofer.** 2003. Impact of insecticide drift on aphids and their parasitoids: residual toxicity, persistence and recolonisation. Agric. Ecosyst. Environ. 94: 265–274.

101. **Latgé, J. P., and B. Papierok.** 1988. Aphid pathogens, pp. 323–335. *In* A. K. Minks and P. Harrewijn [Eds.]. Aphids. Their biology, natural enemies and control. Vol. B. Elsevier, Amsterdam.

102. **Lattin, J. D.** 1989. Bionomics of the Nabidae. Annu. Rev. Entomol. 34: 383–400.

103. **Lattin, J. D.** 1999. Bionomics of the Anthocoridae. Annu. Rev. Entomol. 44: 383–400.

104. **Lattin, J. D.** 2000. Minute pirate bugs (Anthocoridae), pp. 607–637. *In* C. W. Schaefer and A. R. Panizzi [Eds.]. Heteroptera of economic importance. CRC Press, Boca Raton, FL.

105. **Lattin, J. D., A. Chrisie, and M. D. Schwartz.** 1995. Native black grass bugs (*Irbisia-Lapops*) on introduced wheat grasses; commentary and annotated bibliography (Hemiptera: Heteroptera: Miridae). Proc. Entomol. Soc. Wash. 97: 90–111.

106. **Legg, D. E., G. L. Hein, and F. B. Peairs.** 1991. Sampling Russian wheat aphid in the western Great Plains. Colo. St. Univ. Coop. Ext.–Gt. Plains Agric. Counc. Publ. GPAC 138.

107. **Lhaloui, S.** 1995. Biology, host preference, host suitability, and plant resistance studies of the barley stem gall midge and Hessian fly (Diptera: Cecidomyiidae) in Morocco. Ph.D. dissertation, Kansas State University, Manhattan.

108. **Lhaloui, S., M. El Bouhssini, J. H. Hatchett, and K. Starks.** 1987. Grey fly damage to small grains, pp. 42–43. Programme Aridoculture. INRA-CRRA de la Chaouia, Abda et Doukkala. Rapport d'activite, 1986–87.

109. **Lhaloui, S., L. Buschman, M. El Bouhssini, K. Starks, D. Keith, and K. El Houssaini.** 1992. Control of *Mayetiola* species (Diptera: Cecidomyiidae) with carbofuran in bread wheat, durum wheat, and barley with yield loss assessment and its economic analysis. Al Awamia. 77: 55–73.

110. **Luginbill, P.** 1928. The fall armyworm. U.S. Dep. Agric. Tech. Bull. 34.

111. **Luginbill, P., and R. H. Painter.** 1953. May beetles of the United States and Canada. USDA Technical Bulletin 1060. Washington, D. C.

112. **Luginbill, P., and T. D. Urbahns.** 1916. The spike-horned leafminer, an enemy of grains and grasses. U.S. Dep. Agric. Bull. 432.

113. **Macvean, C. M.** 1987. Ecology and management of Mormon cricket, *Anabrus simplex* Haldeman, pp. 116–136. *In* J. L. Capinera [Ed.]. Integrated pest management on rangeland: A shortgrass prairie perspective. Westview Press, Boulder, CO.

114. **Mamluk, O. F., O. Tahhan, R. H. Miller, B. Bayaa, K. M. Makkouk, and S. B. Hanounik.** 1992. A checklist of cereal, food legume and pasture and forage crop diseases and insects in Syria. Arab. J. Plant Prot. 10: 225–266.

115. **Mann, J. A., R. Harrington, N. Carter, and R. T. Plumb.** 1997. Control of aphids and barley yellow dwarf virus in spring-sown cereals. Crop Prot. 16: 81–87.

116. **Mason, C. E., M. E. Rice, D. D. Calvin, J. W. Van Duyn, W. B. Showers, W. D. Hutchison, J. F. Witkowski, R. A. Higgins, D. W. Onstad, and G. P. Dively.** 1996. European corn borer: ecology and management. North Central Regional Extension Publication

References

327. Iowa State University, Ames.
117. **Mathre, D. E. [Ed.]. 1997.** Compendium of barley diseases, 2nd ed. APS (American Phytopathological Society) Press, St. Paul, MN.
118. **McKirdy, S. J., and R. A. C. Jones. 1997.** Effect of sowing time on barley yellow dwarf virus infection in wheat: virus incidence and grain yield losses. Aust. J. Agric. Res. 48: 199–206.
119. **McPherson, J. E., and R. M. McPherson. 2000.** Stink bugs of economic importance in America north of Mexico. CRC Press, Boca Raton, FL.
120. **Medina-Guad, S., L. F. Matorell, and R. B. Robles. 1965.** Notes on the biology and control of the yellow aphid of sugarcane, *Sipha flava* (Forbes) in Puerto Rico, pp. 1307–1320. *In* J. Bague [Ed.]. Proceedings of the 12th Congress of the International Society of Sugar Cane Technologists, 28 March–10 April 1965, San Juan, PR. Executive Committee of the I.S.S.C.T.
121. **Meredith, P. 1970.** Bug damage in wheat. N.Z. Wheat Rev. 11: 49–53.
122. **Mesquita, A. L. M., and L. A. Lacey. 2001.** Interactions among the entomopathogenic fungus, *Paecilomyces fumosoroseus* (Deuteromycotina: Hyphomycetes), the parasitoid, *Aphelinus asychis* (Hymenoptera : Aphelinidae), and their aphid host. Biol. Control. 22: 51–59.
123. **Metcalf, R. L., and R. A. Metcalf. 1993a.** Destructive and useful insects: their habits and control, 5th ed. McGraw-Hill, New York.
124. **Metcalf, R. L., and R. A. Metcalf. 1993b.** Grasshoppers or locusts, pp. 9.4–9.9. *In* R. L. Metcalf and R. A. Metcalf [Eds.]. Destructive and useful insects: their habits and control, 5th ed. McGraw-Hill, New York.
126. **Metcalf, R. L., and R. A. Metcalf. 1993d.** Chinch bug, pp. 9.19–9.23. *In* R. L. Metcalf and R. A. Metcalf [Eds.]. Destructive and useful insects: their habits and control, 5th ed. McGraw-Hill, New York.
127. **Miller, R. H. 1991.** Insect pests of wheat and barley in West Asia and North Africa. ICARDA Technical Manual 9 (rev. 2). ICARDA, Aleppo, Syria.
128. **Miller, R. H. and K. S. Pike. 2002.** Insects in wheat-based systems. FAO Plant Prod. Prot. Ser. 30: 367–393.
129. **Miller, R. H., H. C. Harris, and M. J. Jones. 1994.** Crop rotation effects on populations of *Porphyrophora tritici* (Bodenheimer) (Homoptera: Margarodidae) in barley in northern Syria. Arab. J. Plant. Prot. 12: 79-75.
130. **Miller, R. H., and J. G. Morse (eds.). 1996.** Sunn pests and their control in the Near East. FAO Plant Production and Protection Paper 138. FAO, Rome.
131. **Miller, W. A., G. Koev, and B. R. Mohan. 1997.** Are there risks associated with transgenic resistance to luteoviruses? Plant Dis. 81: 700–710.
132. **Morrill, W. L. 1978.** Emergence of click beetles (Coleoptera: Elateridae) from some Georgia grasslands. Environ. Entomol. 7: 895-986.
133. **Morrill, W. L. 1984.** Wireworms: Control, sampling methodology, and effect on wheat yield in Montana. J. Ga. Entomol. Soc. 19: 67-71.
134. **Morrill, W. L. 1995.** Insect pests of small grains. APS (American Phytopathological Society) Press, St. Paul, MN.
135. **Morrill, W. L., and R. W. Kieckhefer. 1971.** Parasitism of the wheat stem maggot in South Dakota. J. Econ. Entomol. 64: 1129–1131.
136. **Morrill, W. L., and G. D. Kushnak. 1999.** Planting date influence on the wheat stem sawfly (Hymenoptera: Cephidae) in spring wheat. J. Agric. Urban Entomol. 16: 123–128.
137. **Morrill, W. L., D. G. Lester, and A. E. Wrona. 1990.** Factors affecting efficacy of pitfall traps for beetles (Coleoptera: Carabidae and Tenebrionidae). J. Entomol. Sci. 25: 284–293.
138. **Morrill, W. L., J. W. Gabor, D. K. Weaver, G. D. Kushnak, and N. J. Irish. 2000.** Effect of host plant quality on the sex ratio and fitness of female wheat stem sawflies (Hymenoptera: Cephidae). J. Econ. Entomol. 29: 195–199.
139. **Morrill, W. L., D. K. Weaver, and G. D. Johnson. 2001.** Trap strip and field border modification for management of the wheat stem sawfly (Hymenoptera: Cephidae). J. Entomol. Sci. 36: 34–45.
140. **Negron, J. E., and T. J. Riley. 1991.** Seasonal migration and overwintering of the chinch bug (Hemiptera: Lygaeidae) in Louisiana. J. Econ. Entomol. 84: 1681–1685.
141. **Oakley, J. N., and K. F. A. Walters. 1994.** A field evaluation of different criteria for determining the need to treat winter wheat against the grain aphid *Sitobion avenae* and the rose-grain aphid *Metopolophium dirhodum*. Ann. Appl. Biol. 124: 195–211.
142. **Olfert, O., M. K. Mukerji, and J. F. Doane. 1985.** Relationship between infestation levels and yield loss caused by wheat midge, *Sitodiplosis mosellana* (Géhin) (Diptera: Cecidomyiidae), in spring wheat in Saskatchewan. Can. Entomol. 117: 593–598.
143. **Olfert, O., J. F. Doane, and M. P. Braun. 2003.** Establishment of *Platygaster tuberosula*, an introduced parasitoid of the wheat midge *Sitodiplosis mosellana*, in Saskatchewan. Can. Entomol. 135: 303–308.
144. **Olsen, C. E., K. S. Pike, L. Boydston, and D. Allison. 1993.** Keys for identification of apterous viviparae and immatures of six small grain aphids (Homoptera: Aphididae). J. Econ. Entomol. 86: 137–148.
145. **Osborn, H. 1932.** Leaf hoppers injurious to cereal and forage crops. U.S. Dep. Agric. Circ. 241.
146. **Parker, B. L., W. Reid, M. El Bouhssini, S.Gouli, M. Skinner, and S. D. Costa. 2003.** Entomopathogenic fungi of *Eurygaster integriceps* Puton (Hemiptera: Scutelleridae): Collection and characterization for development. Biol. Control 27: 260–272.
147. **Paulian, F., and C. Popov. 1980.** Sunn pest or cereal bug, pp. 69–74. *In* E. Hafliger [Ed.]. Wheat, Documenta Ciba–Geigy. CIBA-GEIGY Basle, Switzerland.
148. **Peairs, F. B. 1998.** Aphids in small grains. Colo. State Univ. Coop. Ext. Fact Sheet 5.565.
149. **Pedigo, L. P. 1996.** Entomology and pest management. Prentice Hall, Upper Saddle, NJ.
150. **Pedigo, L. P., and G. D. Buntin [Eds.]. 1994.** Handbook of sampling methods for arthropods in agriculture. CRC Press, Boca Raton, FL.
151. **Pedigo, L. P., S. H. Hutichins, and L. G. Higley. 1986.** Economic injury levels in theory and practice. Annu. Rev. Entomol. 31: 341–368.
152. **Pfadt, R. E. 1988.** Field guide to common western grasshoppers. USDA–APHIS, Wyo. Agric. Exp. Stn. Bull. 912.
153. **Phillips, W. J., and F. W. Poos. 1940.** The wheat jointworm and its control. U.S. Dep. Agric. Farm. Bull. 1006.
154. **Phillips, W. J., and F. W. Poos. 1953.** The wheat strawworm and its control. U.S. Dep. Agric. Farm. Bull. 1323.
155. **Pike, K., L. Boydston, and D. Allison. 1991.** Winged viviparous female aphid species associated with small grains in North America: key, morphological features and measurements. J. Kans. Entomol. Soc. 63: 559–602.
156. **Pike, K. S., and D. Allison. 1991.** Russian wheat aphid: biology, damage, and management. Pacific Northwest Extension Publication PNW371.
157. **Pike, K. S., and A. L. Antonelli. 1981.** Hessian fly in Washington. Wash. State Univ. Res. Bull. XB0909.
158. **Pike, K. S., P. Starý, T. Miller, D. Allison, L. Boydston, G. Graf, and R. Gillespie. 1997.** Small grain aphid parasitoids (Hymenoptera: Aphelinidae and Aphidiidae) of Washington: distribution, relative abundance, and seasonal occurrence; and key to known

North American species. Environ. Entomol. 26: 1299–1311.

159. **Pivnick, K. A., and E. Labbé. 1993.** Daily patterns of activity of females of the orange wheat blossom midge, *Sitodiplosis mosellana* (Géhin) (Diptera: Cecidomyiidae). Can. Entomol. 125: 725–736.

160. **Poehlinan, I. M. 1985.** Adaptation and distribution, pp. 1–17. *In* D. C. Rasmusson [Ed.]. Barley. Agronomy Monograph 26. American Society of Agronomy, Madison, WI.

161. **Popov, G. B. 1989.** Nymphs of the Sahelian Grasshoppers: an illustrated guide. Chatham, Overseas Development Natural Resources Institute.

162. **Poprawski, T. J., and S. P. Wraight. 1998.** Fungal pathogens of Russian wheat aphid (Homoptera: Aphididae). *In* A response model for an introduced pest—the Russian wheat aphid. Thomas Say Publication Series. Entomological Society of America, Lanham, MD.

163. **Porter, D. R., J. D. Burd, K. A. Shufran, J. A. Webster, and G. L. Teetes. 1997.** Greenbug (Homoptera: Aphididae) biotypes: Selected by resistant cultivars or preadapted opportunists? J. Econ. Entomol. 90: 1055–1065.

164. **Post, R. L., and D. K. McBride. 1966.** Barley thrips biology and control. North Dakota State Extension Circular A-292 (rev.), Fargo.

165. **Quisenberry, S. S., and F. B. Peairs [Eds.]. 1998.** A response model for an introduced pest—the Russian wheat aphid. Thomas Say Publications in Entomology, Entomological Society of America, Lanham, MD.

166. **Rasmusson, D. C. [ed.]. 1985.** Barley. Agronomy Monograph No. 26, American Society of Agronomy, Madison, WI.

167. **Ratcliffe, B. C. 1991.** The scarab beetles of Nebraska. Bulletin of the University of Nebraska State Museum No. 12, Lincoln.

168. **Ratcliffe, R. H., and J. H. Hatchett. 1997.** Biology and genetics of the Hessian fly and resistance in wheat, pp. 47–56. *In* K. Bobdari [Ed.]. New developments in entomology. Research Signpost, Scientific Information Guild, Trivandrum, India.

169. **Ratcliffe, R. H., H. W. Ohm, and F. L. Patterson. 2003.** Hessian fly, pp. 247–260. *In* J. Janick [Ed.]. Breeding wheat for resistance to insects. Plant Breeding Reviews 22. John Wiley & Sons, Hoboken, NJ.

170. **Rice, M. E., and G. E. Wilde. 1988.** Experimental evaluation of predators and parasitoids in suppressing greenbugs (Homoptera: Aphididae) in sorghum and wheat. Environ. Entomol. 17: 836–841.

171. **Riley, C. V., and C. L. Marlatt. 1891.** Wheat and grass saw-flies. Insect Life 4: 168–179.

172. **Runyon, J. B., W. L. Morrill, D. K. Weaver, and P. R. Miller. 2002.** Parasitism of the wheat stem sawfly (Hymenoptera : Cephidae) by *Bracon cephi* and *B. lissogaster* (Hymenoptera: Braconidae) in wheat fields bordering tilled and untilled fallow in Montana. J. Econ. Entomol. 95: 1130–1134.

173. **Schwartz, M. D., and R. G. Foottit. 1992.** Lygus bugs on the prairies: biology, systematics, and distribution. Agric. Can. Res. Br. Tech. Bull. 1992-4E.

174. **Schwartz, M. D., and R. G. Foottit. 1998.** Revision of the Nearctic species of the genus *Lygus* Hahn, with a review of the Palaearctic species (Heteroptera: Miridae). Memoirs on Entomology Vol. 10. International Associated Publishers, Gainesville, FL.

175. **Smith, C. M., S. S. Quisenberry, and F. du Toit. 1999.** The value of conserved wheat germplasm possessing arthropod resistance, pp. 25–49. *In* S. L. Clement and S. S. Quisenberry [Eds.]. Global plant genetic resources for insect resistant crops. CRC Press, Boca Raton, FL.

176. **Smith, R. C., E. G. Kelly, G. A. Dean, H. R. Bryson, and R. L. Parker. 1962.** Insects in Kansas. Kansas State Agricultural Experiment Station Bulletin, Manhattan.

177. **Solie, J. B., S. G. Solomon, Jr., K. P. Self, T. F. Peeper, and J. A. Koscelny. 1991.** Reduced row spacing for improved wheat yields in weed-free and weed-infested fields. Trans. ASAE (Am. Soc. Agric. Eng.) 34: 1654–1660.

178. **Sparks, A. N. 1979.** A review of the biology of the fall armyworm. Fla. Entomol. 2: 82–87.

179. **Spencer, K. A. 1973.** Agromyzidae (Diptera) of Economic Importance, pp. 293–296. Vol. 9, Series Entomologica. W. Junk, The Hague, The Netherlands.

180. **Spencer, K. A., and G. C. Steyskal. 1986.** Manual of the Agromyzidae (Diptera) of the United States, p. 91. U.S. Dep. Agric. Agric. Handb. 638.

181. **Spike, B. P., G. E. Wilde, T. W. Mize, R. J. Wright, and S. D. Danielson. 1994.** Bibliography of the chinch bug *Blissus leucopterus leucopterus* (Say) (Heteroptera: Lygaeidae) since 1888. J. Kansas Entomol. Soc. 67: 116–125.

182. **Starks, K. J., and K. A. Mirkes. 1979.** Yellow sugarcane aphid: plant resistance in cereal crops. J. Econ. Entomol. 72: 486–488.

183. **Starý, P. 1988.** Natural Enemies, 9.1 Parasites, pp. 171–188. In: A. K. Minks and P. Harrewijn [Eds.]. Aphids, their biology, natural enemies and control, Vol. 2B. Elsevier, New York.

184. **Steffey, K. L., M. E. Rice, J. All, D. A. Andow, M. E. Gray, and J. W. Van Duyn [Eds.]. 1999.** Handbook of corn insects. Entomological Society of America, Lanham, MD.

185. **Sunderland, K. D. 1987.** Spiders and cereal aphids in Europe. Bulletin of the Western Palearctic Regional Section IOBC. 10: 82–102.

186. **Symondson, W. O. C., K. D. Sunderland, and M. H. Greenstone. 2002.** Can generalist predators be effective biocontrol agents? Annu. Rev. Entomol. 47: 561–594.

187. **Tashiro, H, C. L. Murdoch, R. W. Straub, and P. J. Vittum. 1977.** Evaluation of insecticides on *Hyperodes* sp., a pest of annual bluegrass turf. J. Econ. Entomol. 70: 729–733.

188. **Thomas, M. B., S. D. Wratten, and N. W. Sotherton. 1991.** Creation of island habitats in farmland to manipulate populations of beneficial arthropods - predator densities and emigration. J. Appl. Ecol. 28: 906–917.

189. **Throne, J. E. 1980.** Bibliography of the seedcorn maggots *Hylemya platura* and *H. florilega* (Diptera: Anthomyiidae). N.Y. State Agric. Exp. Stn. Spec. Rep. 37.

190. **Tippins, H. H., [Ed.]. 1982.** A review of information on the lesser cornstalk borer *Elasmopalpus lignosellus* (Zeller). Ga. Agric. Exp. Stn. Spec. Publ. 17.

191. **Toba, H. H., L. E. O'Keeffe, K. S. Pike, E. A. Perkins, and J. C. Miller. 1985.** Lindane seed treatment for control of wireworms (Coleoptera: Elateridae) on wheat in the Pacific Northwest. Crop Protection 4: 372-380.

192. **Todd, J. W. 1989.** Ecology and behavior of *Nezara viridula*. Annu. Rev. Entomol. 34: 273–292.

193. **Unger, P. W., E. G. Krenzer, Jr., and C. A. Norwood. 1994.** Soil–water conservation, pp. 16–17. *In* B. A. Stewart and W. C. Moldenhauer [Ed.]. Crop Residue Management to Reduce Erosion and Improve Soil Quality. Conservation Research Report 37. USDA-ARS, Sprpingfield, VA.

194. **Varis, A.-L. 1972.** The biology of *Lygus rugulipennis* Popp. (Het. Miridae) and the damage caused by this species in sugarbeet. Ann. Agric. Fenn. 11: 1–56.

195. **Viator, H. P., A. Pantoja, and C. M. Smith. 1983.** Damage to wheat seed quality and yield by the rice stink bug and southern green stink bug (Hemiptera: Pentatomidae). J. Econ. Entomol. 76: 1410–1413.

196. **Voss, T. S., R. W. Kieckhefer, B. W. Fuller, M. J. McLeod, and D. A. Beck. 1997.** Yield losses in maturing spring wheat caused by cereal aphids (Homoptera: Aphididae) under laboratory conditions. J. Econ. Entomol. 90: 1346–1350.

197. **Walkden, H. H. 1950.** Cutworms, armyworms, and related species attacking cereal and forage crops in the Central Great Plains.

References

U.S. Dep. Agric. Circ. 849.

198. **Ward, R. H., and A. J. Keaster. 1977.** Wireworm baiting: Use of solar energy to enhance early detection of *Melanotus depressus*, *M. verberans*, and *Aeolus mellillus* in Midwest cornfields. J. Econ. Entomol. 70: 403-406.

199. **Webster, J. A., and P. Kenkel. 1999.** Economic, environmental, and social benefits of resistance in field crops. Thomas Say Publication, Entomological Society of America, Lanham, MD.

200. **Webster, J. A., C. Inayatullah, M. Hamissou, and K. A. Mirkes. 1994.** Leaf pubescence effects in wheat on yellow sugarcane aphids and greenbugs (Homoptera: Aphididae). J. Econ. Entomol. 87: 231–240.

201. **Wellso, S. G., J. E. Araya, and R. P. Hoxie. 1991.** Wheat pests—arthropod pest management, pp. 137–174. *In* D. Pimentel [Ed.]. CRC handbook of pest management in agriculture, 2nd ed. Vol. III. CRC Press, Boca Raton, FL.

202. **Wheeler, A. G., Jr., and T. J. Henry. 1992.** A synthesis of the Holarctic Miridae (Heteroptera): distribution, biology and origin, with emphasis on North America. Entomological Society of America, Thomas Say Foundation, Vol. 15. Lanham, MD.

203. **Wicks, G. A. 1984.** Integrated systems for control and management of downy brome (*Bromus tectorum*) in cropland. Weed Sci. 32:(Suppl. 1): 26–31.

204. **Wiese, M. V. (Ed.) 1987.** Wheat streak mosaic, pp. 80–81. *In* Compendium of wheat diseases, 2nd ed. American Phytopathological Society Press, St. Paul, MN.

205. **Wilson, H. R. 1980.** European corn borer, *Ostrinia nubilalis* (Lepidoptera: Pyralidae), on wheat. Can. Entomol. 112: 861–863.

Glossary

abdomen. The third or posteriormost of the three major body division of an insect, or the second or posteriormost of the two major body divisions of spiders and mites.

active ingredient. The substance in a pesticide that is responsible for the pesticidal effect. Also known as toxicant.

adventitious roots. Roots that originate from nodes on the stem.

aestivate. To spend the summer (or a warm or dry period) in a dormant condition; opposed to hibernate.

aflatoxin. Any of several carcinogenic mycotoxins that are produced by *Aspergillus* molds, especially in stored agricultural crops.

alate. The winged form of an aphid.

allomone. A chemical transmitted between species that benefits the releaser species but not the receiver species, such as a repellent or toxicant.

anal plate. A hardened "plate" on the dorsal surface of the last abdominal segment.

anholocycly. Year-round parthenogenetic reproduction in insects such as aphids.

annulet. A partial, dorsal subdivision of a body segment formed by transverse creases in the cuticle.

antennae. In larval and adult stages of an insect, the paired segmented appendages located on each side of the head that function as sense organs.

antennal club. The enlarged distal segments of a clubbed antenna.

anterior. Toward the front, as opposed to posterior; in front of.

anthesis. Flowering and pollination period of cereal grains.

anthers. The parts of male flowers that produce pollen.

apex. That part of any joint or segment opposite the base or point of attachment to the body.

apical. At, near, or pertaining to the end, tip, or outermost part.

apterous. Wingless.

arthropods. Invertebrate animals with jointed appendages; members of the phylum Arthropoda.

augmentation. A method of biological control that enhances effectiveness of a biological control agent by mass release, genetic improvement, or use of semiochemicals to alter behavior.

avirulence. Inability to live on and injury plants.

awn. Bristle-like structure at the apex of the outer bract of a cereal flower.

basal. At or near the point of attachment to the main body.

beneficial. A useful insect, often one that is a predator or parasitoid of a harmful insect.

bifuraction. Forked or dividing into two.

biological control. The actions of parasites, parasitoids, predators, and pathogens that lower host population densities; and the study and use of parasites, parasitoids, predators, and pathogens for the regulation of host population densities.

biotype. A population of individuals that differs from the rest of the species by some criteria other than morphology, such as biochemical, physiological or behavioral properties.

bivoltine. Having two generations per year.

brachypterous. Having short wings that do not cover the abdomen.

cantharidin. A toxin produced by blister beetles

caterpillar. Usually refers to the larva of a moth or butterfly.

cauda. The tail or any process resembling a trail; the pointed end of the abdomen in aphids.

cephalothorax. In spiders and other arachnids, the body region consisting of head and thoracic segments.

cercus (pl. cerci). An appendage, normally paired, segmented and usually slender, extending from the end of the abdomen.

cereal. A small-grained grass grown for its edible seed.

chaff. Non-seed portion of mature cereal seed head or spike.

chelicerae. A major element in the mouthparts of spiders and related arthropods; in the form of fangs, pincers, or piercing organs.

chitin. A colorless nitrogenous polysaccharide occurring in the cuticle of arthropods.

chlorosis. Yellowing or blanching of green plants.

chorion. The outermost protective layer around an insect egg; the 'shell'.

chromosome. Replicating rob-like body of DNA in the cell nuclei; group of genes.

class. A subdivision of a phylum or subphylum, containing a group of related orders.

clavate. Thickening gradually toward the tip.

Glossary

cocoon. The silken or fibrous covering constructed by the larva for protection during its pupal period.

coleoptile. The sheath enclosing the apical meristem and leaf primordia of the grass embryo; often interpreted as the first leaf.

collar. The junction between the leaf blade and leaf sheath in grasses.

complete metamorphosis. Metamorphosis of an insect that includes four stages of development: egg, larva, pupa, adult.

conservation. A method of biological control in which the environment is modified to reduce or eliminate conditions unfavorable to natural enemies or to promote population growth, recruitment and performance of natural enemies.

cornicles. A pair of erect tubules located on the end of the abdomen of aphids that secrete a defensive liquid and sometimes an alarm pheromone.

coxa (plural **coxae**). The basal segment of the insect's leg, by which it is articulated to the body.

crochets. Curved spines or hooks on the underside of the prolegs of caterpillars.

cuticle. The outer covering of an insect formed by a layer of chitin.

crown. Region at the base of the stem from which tillers arise.

cytoplasmic polyhedrosis virus. Virus in the family Reoviridae, which are icosahedral, double-stranded RNA viruses.

culm. Stem of a grass plant.

cultivar. A cultivated variety.

damage. A measurable reduction in plant utility caused by insect injury resulting in loss of plant growth, seed yield or quality.

dead heart. Wilting, discoloration, and death of whorl leaves when stem feeding (boring) activity of insects cuts off water and nutrient flow.

degree day. See heat unit.

developmental threshold. The minimum temperature required for development.

deutonymph. The third instar of a mite.

diapause. A period of arrested development and reduced metabolic rate during which growth, differentiation, and metamorphosis cease; a period of dormancy not immediately referable to adverse environmental conditions.

distal. That part of a segment or appendage farthest from the body.

diurnal. Active during the daytime.

DNA. Deoxyribonucleic acid; chemical encoding an organism's genetic composition.

dormancy. A state of reduced physiological activity.

dorsal. Pertaining to the upper surface, back, or top side.

dorsum. The upper surface, back, or top side.

double cropping. Growing two crops in the same field in one year, such as winter wheat followed by soybean.

ear. Spike or grain head of a cereal plant.

ecdysis. Molting; the process of shedding the exoskeleton.

eclosion. Emergence of the adult from the pupa; hatching from the egg.

economic injury level. The pest density or level of injury that will cause yield loss equal to the cost of controlling the pest.

economic threshold. The pest density at which management action should be taken to prevent an increasing pest population from reaching the economic injury level.

ecosystem. A living community and its nonliving environment.

ectoparasitoid. A parasitoid in which immature stages feed and develop outside the host.

elytra (singular **elytron**). Thickened, leathery, or horny front wings, primarily in Coleoptera (beetles) and Dermaptera (earwigs).

emergence. The act of the adult insect leaving the pupal case or the last nymphal skin.

endoparasitoid. A parasitoid in which immature stages feed and develop inside the host.

endotoxin. A poisonous substance present in bacteria but separable from the cell body only on its disintegration.

entomogenous. Microorganisms growing in or on the bodies of insects; refers to a parasitic relationship.

entomopathogen. Organism that is pathogenic to arthropods.

entomophagous. Insect-eating.

epidemiology. Study of disease initiation, development, and spread.

epidermis. The cellular layer of the body wall, which secretes the cuticle.

epizootic. Outbreak of a disease that suddenly and temporarily affects an animal population.

exoskeleton. The skeleton or supporting structure on the outside of the body.

exotic. Introduced from a foreign country.

feces. Excrement, the material passed from the alimentary tract through the anus.

femur. The third leg segment, located between the trochanter and the tibia.

feral. Wild, including having escaped from domestication and become wild.

filiform. Threadlike; slender and of equal diameter.

flag leaf. Uppermost leaf in reproductive stage cereal plants.

forewing. The anterior or front pair of wings.

formulation. A preparation containing a pesticide in a form suitable for practical use.

Glossary

frass. Solid larval insect excrement.
fungicide. Chemical that kills or limits growth of fungi.
gall. Abnormal plant growth caused by insect feeding
gene. Smallest functional unit of genetic material on a chromosome.
generation. A group of offspring of the same species that develop in approximately the same time frame.
genitalia. Sexual organs and associated structures.
genome. Group of chromosomes.
genus (plural **genera**). A group of closely related species; the first name in a binomial or trinomial scientific name. The genus name is Latinized, capitalized and, when printed, italicized.
growing point. A part of the plant body at which cell division is localized, generally terminal and composed of meristematic cells.
grub. Thick-bodied larva with a well developed head an thoracic legs, without abdominal prolegs, and usually sluggish. Term used to describe some beetle larvae.
head. The anterior body region, which bears the eyes, antennae, and mouthparts.
head capsule. The combined sclerites of the head that form a hard, compact case.
heat unit. An accumulation of degrees above some threshold temperature during a 24-hour period; a measure of physiological time for cool-blooded organisms, like insects; also called degree-day.
habitat. The natural abode of a plant or animal, especially the particular location where it normally lives.
hemimetabolous. Insects with incomplete metamorphosis, but nymphs and adult usually occupy different habitats and differ in appearance.
herbicide. Chemical that kills unwanted plant or limits their growth.
herbivorous. Feeding on plants.
heteroecious. Alternating between two unrelated host plants during part of the annual life cycle as in aphids and other insects.
hibernate. To pass the winter in a dormant state.
hindgut. The posterior region of the digestive tract, between the midgut and anus.
holocycly. Insects such as aphids having males and females reproducing sexually during part of the year and females reproducing by parthenogenesis the rest of the year.
holometabolous. Having a complete transformation, with egg, larval, pupal, and adult stages distinctly separated. Undergoing complete metamorphosis.
honeydew. A liquid rich in sugar excreted from the anus of an aphid.
hybrid. Offspring of two individuals of different genetic character.
hyperparasite. An insect that parasitizes another parasitic insect.
importation or classical biological control. A method of biological control that introduces and seeks to establish exotic natural enemies.
incomplete metamorphosis. Undergoing development that is gradual and lacks a sharp separation into larval, pupal, and adult stages (larval stages [nymphs] are often similar to adults in appearance and feeding behavior).
indigenous. Native to an area.
inert ingredient. Any material in a pesticide formulation having no pesticidal action itself.
infection. The introduction or entry of a pathogenic microorganism into a susceptible host, resulting in the presence of the microorganism within the body of the host, whether or not this causes detectable pathological effects.
injury. Pest activity causing a change in plant physiological processes.
insecticide. Chemical used to control or kill insects.
instar. The stage of an insect between successive molts, the first instar being the stage between hatching and the first molt.
intercropping. The growing together of two or more crops in the same field at the same time.
integrated pest management (IPM). A system of economically and environmentally sound practices to reduce the deleterious impact of pest activities; frequently associated with the use of multiple management tactics (e.g., pesticides, cultural control, mechanical control, biological control).
internode. Culm or stem area between two consecutive nodes.
joint. Node of grass stem
jointing. Cereal plant growth stage of rapid stem/culm elongation.
kairomone. A chemical transmitted between species that benefits the receiver species but not the releaser species, such as an attractant or excitant.
labial palps. A pair of small feeler-like structures arising from the labium (part of the mouthparts of an insect).
labium. One of the mouthpart structures, the lower lip.
labrum. The upper lip, lyping just below the face.
larva. The immature stage between egg and pupa in insects with complete metamorphosis.
leaf sheath. The extension of the leaf base covering the stem.
lodging. Plants or culms falling over; stems not upright. Usually in mature plants with some type of injury and after high winds.
maggot. A legless larva without a well-developed head capsule. A term used to describe some fly larvae.

Glossary

mandibles. Part of the mouthparts of insects, usually used to chew food, but sometimes modified into structures for piercing or sucking.

maxillae. The hind or second set of jaws behind the mandibles.

median. In the middle; along the midline of the body.

meristem. The undifferentiated plant tissue from which new cells arise.

mesocotyl. The internode between the seed and the coleoptile in grass plants.

mesophyll. The leaf substance lying between the upper and lower epidermis.

mesopleuron. The sclerites along the side of the mesothorax.

mesothorax. The middle or second segment of the thorax.

metamorphosis. The process of changes that an insect passes through during its growth from egg to adult.

midrib. Central thickened vein of grass leaves.

migration. Any cyclical movement (usually annual) that occurs during the life history of an animal at definite intervals and always includes a return trip from where they began.

molt. To cast off or shed the outer skin. (exoskeleton) at certain intervals to accommodate growth of the body.

monocot. A subdivision of flowering plants whose members possess one embryonic seed leaf, or cotyledon; includes small grains.

moth. An adult insect (Lepidoptera) with two pairs of scale-covered wings and variously shaped (but never clubbed) antennae.

multivoltine. Having many generations per year.

natural control. The actions of abiotic factors (such as weather) and biotic factors (such as natural enemies) that lower population densities of an organism.

necrosis. Usually localized death and discoloration of living tissue.

no tillage. A system of crop production where crops are planted without cultivation into previous crop residue.

nocturnal. Active at night.

node. Knob or joint of a stem at which leaves arise.

nymph. The immature stage in insects with incomplete metamorphosis.

omnivorous. Feeding on a variety of substances of both animal and vegetable origin.

overwinter. To survive the winter.

oviparae. Mating, egg-laying female.

oviposit. To lay or deposit eggs.

oviposition. Egg laying.

ovipositor. The tubular structure through which female insects deposit their eggs; also a wasp's "stinger".

parasitoid. A species that parasitizes another species, in which a relatively small number of immature stages slowly consume the host, and adult stages actively search for hosts; also called a protelean parasite. These include Tachinidae and parasitic Hymenoptera.

parthenogenesis. Asexual reproduction; eggs undergo full development without being fertilized.

pathogen. An organism that causes disease. A microorganism capable of producing disease under normal conditions of host resistance and rarely living in close association with the host without producing disease. Any microorganism, virus, substance, or factor causing disease.

pathogenic. Causing disease.

paurometabolous. Insect with incomplete metamorphosis with winged adults and nymphs living in the same habitat.

perennial. Living for more than two years or growing seasons.

pedicel. The 'waist' of an ant and wasp, made up of either one or two segments of the base of the abdomen; the second segment of a jointed antenna from the base outward.

pericarp. The fruit wall that develops from the mature ovary wall.

persistent. Type of plant virus that remains active and infectious within insect vectors for long periods.

phenology, phenological. Temporal and seasonal pattern of life-history events in plants and animals.

pheromone. A chemical substance secreted by an animal that influences the behavior of other individuals of the same species.

phloem. Food-conducting tissue of plant vascular system.

photosynthesis. Plant process of manufacturing carbohydrates from carbon dioxide and water, using light energy.

phytophagous. Feeding upon plants.

phytotoxic. Toxic to some plants.

pollen. Microspores produced by the male flower; usually appears as a fine dust.

polymorphic. Having many different forms or sizes.

polyphagous. Eating many kinds of foods.

posterior. Toward the rear, as opposed to anterior.

predator. An animal that attacks and feeds on other animals (its prey).

prepupa. A quiescent instar between the active larval period and pupa stage.

proboscis. Any extended, beaklike mouth structure.

prolegs. Fleshy, nonsegmented abdominal walking appendages of some insect larvae.

pronotum. The dorsal 'plate' of the prothorax.

prothoracic shield. Another term for pronotum.

protonymph. The second instar of a mite.

pubescence. Short, fine hairs.

pupa. The stage between the larva and the adult in

insects with complete metamorphosis; a nonfeeding and usually an inactive stage.

puparium. A case formed by the hardening of the last larval skin, in which the pupa is formed (in Diptera).

raster. Complex of definitely arranged bare areas, hairs, and spines on the ventral surface of the last abdominal segment, in front of the anus of some beetle larvae.

resistance (arthropod). The ability of pests to avoid or mitigate the toxic effects of pesticides, whether by physical, physiological, or behavioral means.

resistance (plant). Property of plant hosts that impedes or prevent arthropod attack, oviposition and/or development.

RNA. Ribonucleic acid.

saprophytic. Living on dead or decaying matter.

scale. A flat, unicellular outgrowth of the insect body wall, of various shapes.

sclerite. A hardened body wall plate bounded by sutures or membranous areas.

sclerotized. Hardened; pertaining to portions of the insect integument that are hardened in definite areas by formation of substances other than chitin.

semiochemical. Any chemical involved in communications among organisms.

senescence. The process or state or growing old.

sequential sampling. A procedure whereby management decisions and the need to take more samples to reach a decision are based on the cumulative results of a sequence of samples.

serrate. Toothed along the edge like a saw.

sessile. Attached or fastened; not free to move about.

setae (singular **seta**). Bristles; stiff hairs.

silks. Long slender tubes produced by ovules on the corn ear; the silks receive pollen and conduct the genetic contests of the pollen grain to the female flower on the cob.

siphunculus (plural **siphunculi**). Tubular abdominal structure of aphids from which defensive compounds and sometimes alarm pheromones are expelled. Also called cornicle.

species. The smallest taxonomic group; a population that has a defined range and can exchange genes.

spike. Inflorescence with florets or spikelets on an axis; cereal plant ear or grain head.

spiracle. An external opening of the tracheal system; a breathing pore.

spores. Reproductive body of the lower plants.

spring wheat. Wheat sown in spring to mature in the same growing season.

stadium (plural **stadia**). The period between molts of a developing arthropod.

stage. An insect's developmental status (e.g., the egg stage).

sternite. The ventral plate of an abdominal segment.

sternum. The entire ventral division of any segment; the underside of the insect thorax.

systemic insecticide. An insecticide that is absorbed and translocated (moved) to other tissues within a plant.

tarsal claw. Structure found at the end of the tarsus.

tarsus (plural **tarsi**). Segmented part of the leg immediately beyond the tibia, sometimes consisting of one or more segments or subdivisions.

taxon (plural **taxa**). A taxonomic category, such as phylum, class, order, family, genus, species.

terminal process. In aphids the tip or end of the antennae.

thorax. The body region of an insect between the head and the abdomen that bears the legs and wings.

tibia. The fourth segment of the leg, between the femur and the tarsus.

tiller. Shoot, culm, or stalk arising from the crown of a grass plant.

tolerant. Sustaining insect injury without serious damage or yield loss.

toxin. A poisonous substance.

transgenic organism. An organism containing certain genes from another species that is artificially produced, for example, by injecting foreign DNA into the nuclei of egg cells or early embryos.

triticale. A hybrid cereal from a cross of wheat and rye.

tubercle. A small knot-like or rounded protuberance.

ubiquitous. Omnipresent; occurs everywhere.

univoltine. Having one generation per year.

vector. An organism, such as an insect, that transmits a pathogen.

ventral. Pertaining to the underside of the body.

vestigial. Small or degenerate; the remains of a previously functional part.

viviparous. Bearing live young.

virulence. Capacity to injury plants.

viruliferous. Virus-laden.

white head. Bleached, dry immature spike.

whorl. A circle of leaves in a cereal plant.

wing pads. The underdeveloped wings of nymphs of insects with incomplete metamorphosis (e.g., Hemiptera), that appear on the thorax as two lateral, flat structures.

winter wheat. Wheat sown in the fall that matures the next spring.

xylem. The principal strengthening and water-conducting tissue of branches, stems, and roots.

Sources of Local Information

United States, by State

A

Department of Entomology & Plant Pathology
301 Funchess Hall
Auburn University
Auburn University, AL 36849-5413
(334) 844-5006
FAX: (334) 844-5005
http://www.ag.auburn.edu/dept/ent/ent.html

Department of Entomology
410 Forbes Building
University of Arizona
Tucson, AZ 85721
(520) 621-1151
FAX: (520) 621-1150
http://ag.arizona.edu/ENTO/entohome.html

Department of Entomology
321 Agriculture Building
University of Arkansas
Fayetteville, AR 72701
(501) 575-2451
FAX: (501) 575-2452
http://www.uark.edu/depts/entomolo/

C

Department of Environmental Science,
Policy, and Management
145 Mulford Hall #3114
University of California
Berkeley, CA 94720
(510) 643-2626
FAX: (510) 642-6632
http://www.cnr.berkeley.edu/departments/espm/

Department of Entomology
1 Shields Ave.
University of California
Davis, CA 95616
(530) 752-0475
FAX: (530) 752-1537
http://www.aes.ucdavis.edu/ex/departments/Dep_entomology.htm

Department of Entomology
University of California
Riverside, CA 92521-0314
(909) 787-3718
FAX: (909) 787-3086
http://cnas.ucr.edu/~ento/

Department of Bioagricultural Sciences and Pest Management
Colorado State University
Fort Collins, CO 80523-1177
(970) 491-5261
FAX: (970) 491-3862
http://www.colostate.edu/Depts/Entomology/ent.html

Department of Ecology and Evolutionary Biology
75 N Eagleville Rd., U-43
University of Connecticut
Storrs, CT 06269-3043
(860) 486-4322
FAX: (860) 486-6364
http://www.biology.uconn.edu

D

Department of Entomology and Applied Ecology
University of Delaware
Newark, DE 19717-1303
(302) 831-2526
FAX: (302) 831-3651
http://ag.udel.edu/departments/ento

Department of Entomology
National Museum of Natural History
Smithsonian Institution
Washington, DC 20560-0105
(202) 357-2078
FAX: (202) 786-2894
http://entomology.si.edu/textmenu.html

F–G

Fort Lauderdale Research and Education Center
3205 College Avenue
University of Florida
Fort Lauderdale, FL 33314-7799
(954) 475-8990
FAX: (954) 475-4125
http://www.ftld.ufl.edu

Department of Entomology and Nematology
P.O. Box 110620
University of Florida
Gainesville, FL 32611-0620
(352) 392-1901
FAX: (352) 392-0190
http://www.ifas.ufl.edu/~entweb/entomolo.htm

Center for Studies in Entomology
Florida A & M University
Tallahassee, FL 32307-4100
(850) 599-3912
FAX: (850) 561-2221
http://www.famu.org/ent/index.html

Department of Entomology
413 Biological Sciences Bldg.
University of Georgia
Athens, GA 30602-2603
(706) 542-2816
FAX: (706) 542-2279
http://entomology.ent.uga.edu

Department of Entomology
Coastal Plain Experiment Station
University of Georgia
Tifton, GA 31793
(912) 386-3374
FAX: (912) 386-3086
http://sacs.cpes.peachnet.edu/entomolo.htm

H–I

Department of Entomology
Gilmore 310, 3050 Maile Way
University of Hawaii at Manoa
Honolulu, HI 96822-2271
(808) 956-7076

Sources of Local Information

FAX: (808) 956-2428
http://www2.ctahr.hawaii.edu/ento

Department of Plant, Soil, and Entomological Sciences
University of Idaho
Moscow, ID 83844-2339
(208) 885-6276
FAX: (208) 885-7760
http://www.uidaho.edu/pses/

Department of Entomology
320 Morrill Hall
505 S. Goodwin Avenue
University of Illinois
Urbana, IL 61801
(217) 333-2910
FAX: (217) 244-3499
http://www.life.uiuc.edu/entomology/home.html

Department of Entomology
1158 Smith Hall
Purdue University
West Lafayette, IN 47905-1158
(765) 494-4554
FAX: (765) 494-0535
http://www.entm.purdue.edu

Department of Entomology
Insectary Bldg
Iowa State University
Ames, IA 50011-3222
(515) 294-7400
FAX: (515) 294-5957
http://www.ent.iastate.edu/

K–L

Department of Entomology
Haworth Hall
University of Kansas
Lawrence, KS 66045-2106
(785) 864-4301
FAX: (785) 864-5321
http://www.ukans.edu/~entomol/

Department of Entomology
123 Waters Hall
Kansas State University
Manhattan, KS 66506-4004
(785) 532-6154
FAX: (785) 532-6232
http://www.oznet.ksu.edu/entomology

Department of Entomology
S-225 Agricultural Science Center North
University of Kentucky
Lexington, KY 40546-0091
(606) 257-7450
FAX: (606) 323-1120
www.uky.edu/Agriculture/Entomology/enthp.htm

Department of Entomology
402 Life Sciences Bldg.
Louisiana State University
Baton Rouge, LA 70803-1710
(225) 388-1634
FAX: (225) 388-1643
http://www.coa.lsu.edu/entom/entom.html

M

Maine Agricultural & Forestry Experiment Station
5782 Winslow Hall
University of Maine
Orono, ME 04469
(207) 581-3202
http://www.umext.maine.edu

Department of Entomology
4112 Plant Sciences Bldg.
University of Maryland
College Park, MD 20742-4454
(301) 405-3911
FAX: (301) 314-9290
http://www.entm.umd.edu/

Department of Entomology
102 Fernald Hall
270 Stockbridge Road
University of Massachusetts
Amherst, MA 01003-2410
(413) 545-2283
FAX: (413) 545-2115
http://www.umass.edu/ent/

Museum of Comparative Zoology
26 Oxford St.
Harvard University
Cambridge, MA 02138
(617) 495-3045
FAX: (617) 495-5667
http://www.mcz.harvard.edu

Department of Entomology
243 Natural Science Building
Michigan State University
East Lansing, MI 48824-1115
(517) 355-4665
FAX: (517) 353-4354
http://www.ent.msu.edu/dept/

Department of Entomology
219 Hodson Hall
University of Minnesota
St. Paul, MN 55108
(612) 624-3636
FAX: (612) 625-5299
http://ent.agri.umn.edu/

Department of Entomology and Plant Pathology
Box 9775
Mississippi State University
Mississippi State, MS 39762-9775
(601) 325-2085
FAX: (601) 325-8837
http://msstate.edu/entomology/ENTPLP.html

Department of Entomology
1-87 Agriculture Building
University of Missouri
Columbia, MO 65211
(573) 882-7894
FAX: (573) 882-1469
http://cafnr.missouri.edu/plant-Science/entomology/index.stm

MSU Entomology Group
Montana State University
119 Linfield Hall
P.O. Box 172900
Bozeman, MT 59717-2900
Tel: (406) 994-5690
FAX: (406) 994-5885
http://entomology.montana.edu/

N

Department of Entomology
202 Plant Industry Building
University of Nebraska
Lincoln, NE 68583-0816
(402) 472-2123
FAX: (402) 472-4687
http://www.ianr.unl.edu/ianr/entomol/entdept.htm

Sources of Local Information

Department of Entomology
Box 454012
University of Nevada
Las Vegas, NV 89154
(702) 895-1403
FAX: (702) 895-3094
http://hrcweb.lv-hrc.nevada.edu/mbm/ento.html

Department of Entomology
P. O. Box 231
Rutgers University
New Brunswick, NJ 08903-0231
(732) 932-9459
FAX: (732) 932-7229
http://www.rci.rutgers.edu/~insects/

Department of Entomology, Plant Pathology, and Weed Science
Box 30003, Dept. 3BE
New Mexico State University
Las Cruces, NM 88003-0003
(505) 646-3225
FAX: (505) 646-8087
http://taipan.nmsu.edu/eppws/eppws.html

Department of Entomology
2130 Comstock Hall
Cornell University
Ithaca, NY 14853
(607) 255-3253
FAX: (607) 255-0939
http://www.cals.cornell.edu/dept/entomology/

Department of Environmental and Forest Biology
133 Illick Hall
1 Forestry Drive
State University of New York
Syracuse, NY 13210-2788
(315) 470-6743
FAX: (315) 470-6934
http://www.esf.edu/faculty/efb/

Department of Entomology
North Carolina State University
2301-A Gardner Hall
Raleigh, NC 27695-7613
(919) 515-8888
FAX: (919) 515-7746
http://www.cals.ncsu.edu/entomology/

Department of Entomology
202 Hultz Hall
Box 5346, University Station
North Dakota State University
Fargo, ND 58105-5346
(701) 231-7908
FAX: (701) 231-8557
http://www.ndsu.nodak.edu/entomology/

O

Department of Entomology
1735 Neil Ave.
Ohio State University
Columbus, OH 43210
(614) 292-8209
FAX: (614) 292-2180
http://iris.biosci.ohio-state.edu/ouent/

Department of Entomology and Plant Pathology
127 Noble Research Center
Oklahoma State University
Stillwater, OK 74078
(405) 774-5527
FAX: (405) 744-6039
http://www.ento.okstate.edu/

Department of Entomology
2046 Cordley Hall
Oregon State University
Corvallis, OR 97331-2709
(541) 737-4733
FAX: (541) 737-3643
http://www.ent.orst.edu/entomology

P–R

Department of Entomology
501 Agricultural Sciences and Industries Building
Pennsylvania State University
University Park, PA 16802
(814) 865-1895
FAX: (814) 865-3048
http://www.ento.psu.edu/

Crop Protection Department
P.O. Box 9030
University of Puerto Rico
Mayaquez, PR 00681-9030
(787) 265-3859
FAX: (787) 265-0860
http://www.uprm.edu/wciag/cprotgen.htm

Department of Plant Sciences
University of Rhode Island
Kingston, RI 02881
(401) 874-2924
FAX: (401) 874-5296
http://www.uri.edu/cels/pls/

S

Department of Entomology
114 Long Hall
Box 340365
Clemson University
Clemson, SC 29634-0365
(864) 656-3111
FAX: (864) 656-5065
http://entweb.clemson.edu/

Department of Plant Science
Box 2207-A
South Dakota State University
Brookings, SD 57007
(605) 688-5123
FAX: (605) 688-4602
http://www.sdstate.edu/~wpls/
http/pscihome.html

T–U

Entomology and Plant Pathology Department
P.O. Box 1071
University of Tennessee
Knoxville, TN 37901-1071
(865) 974-7135
FAX: (865) 974-4744
http://eppserver.ag.utk.edu

Department of Entomology
Texas A & M University
412 Heep Center
College Station, TX 77843-2475
(979) 845-2516
FAX: (979) 845-6305
http://entowww.tamu.edu/

Department of Plant and Soil Science
Box 42122
Texas Tech University
Lubbock, TX 79409-2122
(806) 742-2838
FAX: (806) 742-0775
http://www.pssc.ttu.edu

Sources of Local Information

Department of Biology
5305 Old Main Hill
Utah State University
Logan, UT 84322-5305
(435) 797-2485
FAX: (435) 797-1575
http://www.biology.usu.edu/

Department of Biology
257 South 1400 East
University of Utah
Salt Lake City, UT 84112-0840
(801) 581-5636
FAX: (801) 581-4668
http://www.biology.utah.edu

V–W

Department of Biology
University of Vermont
Burlington, VT 05405
(802) 238-9765
FAX: (802) 656-2914
http://www.uvm.edu/~biology/

Department of Entomology
Virginia Polytechnic Institute and State University
Blacksburg, VA 24061-0319
(540) 231-6341
FAX: (540) 231-9131
http://www.ento.vt.edu/

Department of Entomology
P.O. Box 646382
Washington State University
Pullman, WA 99164-6382
(509) 335-5504
FAX: (509) 335-1009
http://entomology.wsu.edu/

Division of Plant and Soil Sciences
P.O. Box 6108
West Virginia University
Morgantown, WV 26506-6108
(304) 293-4817 (Division)
(304) 293-6023 (Entomology)
FAX: (304) 293-3740
http://www.caf.wvu.edu/plsc/index.html

Department of Entomology
237 Russell Labs, 1630 Linden Dr.
University of Wisconsin
Madison, WI 53706
(608) 262-3227
http://www.entomology.wisc.edu/

Department of Renewable Resources
P.O. Box 3354
University of Wyoming
Laramie, WY 82071-3354
(307) 766-3103
FAX: (307) 766-3379
http://soils.uwyo.edu/

Canada

Department of Biological Sciences
CW 405 Biological Sciences Bldg.
University of Alberta
Edmonton, AB
Canada T6G 2E9
(780) 492-3308
FAX: (780) 492-9234
http://www.biology.ualberta.ca/

Department of Biological Sciences
8888 University Dr.
Simon Fraser University
Burnaby, BC
Canada V5A 1S6
(604) 291-4475
FAX: (604) 291-3496
http://www.sfu.ca/biology/homepage.html

Department of Entomology
University of Manitoba
Winnipeg, MB
Canada R3T 2N2
(204) 474-9257
FAX: (204) 474-7628
http://www.umanitoba.ca/faculties/afs/entomology/

Department of Biology
Memorial University
St. Johns, NF
Canada A1B 3X9
(709) 737-7497
FAX: (709) 737-3018
http://www.mun.ca/biology/

Department of Environmental Biology
University of Guelph
Guelph, ON
Canada N1G 2W1
(519) 824-4120
FAX: (519) 837-0442
http://www.oac.uoguelph.ca/env

Department of Zoology
University of Western Ontario
London, ON
Canada N6A 5B7
(519) 679-2111
FAX: (519) 661-2014
http://www.uwo.ca/zoo/

Department of Biology
1125 Colonel By Drive
Carleton University
Ottawa, ON
Canada K1S 5B6
(613) 520-3515
FAX: (613) 520-2569
http://www.carleton.ca/biology/

Faculty of Forestry
Earth Sciences Centre
33 Wilcocks St.
University of Toronto
Toronto, ON
Canada M5S.3B3
(416) 978-5751
FAX: (416) 978-3834
http://www.utoronto.ca/forest/termite/termite.htm

Department of Zoology
St. George Campus
25 Kings College Circle
University of Toronto
Toronto, ON
Canada M5S 1A3
(416) 978-2011
FAX: (416) 978-8532
http://www.toronto.com/E/V/TORON/0020/16/68/cs1.html

Lyman Entomological Museum and Research Laboratory
Macdonald Campus of McGill University
Ste. Anne de Bellevue, PQ
Canada H9X 3V9
(514) 398-7914
FAX: (514) 398-7990
http://www.agrenv.mcgill.ca/facility/lyman.htm

Sources of Local Information

Department of Biology
112 Science Place
University of Saskatchewan
Saskatoon, SK
Canada S7N 5E2
(306) 966-4399
FAX: (306) 966-4461
http://www.usask.ca/biology

Mexico

Esp. de Entomologia y Acarologia
Instituto de Fitosanidad
Colegio de Postgraduados
C.P. 56230
Montecillo, Texcoco, Edo. de Mexico
Mexico
(595) 1-15-80 and 1-02-20
FAX: (595) 1-15-80 and 1-02-20

Graduados en Agricultura
**Instituto Tecnologico y de Estudios
    Superiores de Monterrey**
Suc. Correos "J"
Monterrey, N.L. 64849
Mexico
(8) 358-2000 ext. 5190
FAX: (8) 359-9206
http://www.sistema.itesm.mx/

Facultad de Ciencias Biologicas
Universidad Autonoma De Nuevo
    Leon
Cd. Universitaria A.P. F-16
San. Nicolas de Los Garza
Nuevo Leon
C.P. 66450
Mexico
(8) 352-4245 and –6380
http://www.dsi.uanl.mx/

Index

A
Acari, 26, 75, 77, 78, 79
Acer tosichella
Acrididae, 26, 57, 91
Adelphocoris, 66
American wheat striate mosaic virus, 21, 63
Anabrus simplex, 65
Anthocoridae, 98, 103
Aphididae, 26, 37, 38, 39, 40, 42, 43
aphids, 11, 13, 16, 18, 19, 20, 22, 23, 37, 39, 40, 42, 43, 44, 45, 93, 94, 95, 97, 98, 100
Araneae, 99
armyworms, 13, 46, 47, 48, 94

B
bait station, 85
Banks grass mite, 25, 26, 35, 78
barley, 1, 2, 3, 6, 7, 9, 10, 11, 17, 19, 20, 21, 22, 23, 37, 38, 39, 40, 42, 43, 44, 45, 50, 51, 53, 59, 67, 68, 70, 72, 73, 75, 77, 79, 82, 86, 87, 88, 89, 90
barley shoot fly, 67, 90
barley stem gall midge, 89
barley yellow dwarf virus, 11, 20, 38
barley yellow streak mosaic virus, 77
Beauveria bassiana, 93, 94
billbugs, 49
biological control, 1, 18, 19, 41, 60, 75, 93, 94, 95, 96, 97, 98, 100
biotype, 17, 41, 43, 61, 62
bird cherry-oat aphid, 18, 19, 20, 40, 45
black chaff, 22, 40
black fly, 90, 91
Blissus leucopterus, 52, 105
blister beetles, 50, 107
brown stink bug, 70
brown wheat mite, 77

C
cantharidin, 50
Carabidae, 58, 87, 98
Cecidomyiidae, 26, 58, 73, 89
Cephus cinctus, 80
cereal aphid, 38, 39, 93, 97
cereal leaf beetle, 18, 19, 40, 50, 51, 52, 94, 95, 96
cereal yellow dwarf virus, 20
Cerodontha, 63
chinch bug, 52, 53, 93
chromosome, 18
Chrysopidae, 97
Cicadellidae, 63
Coccinella septempunctata, 96
Coccinellidae, 96
Coleomegilla maculata, 96
convergent lady beetle, 96
corn leaf aphid, 20, 38, 39
crop rotation, 7, 14, 16, 23, 73, 80, 83, 88, 91, 99
Ctenicera glauca, 84
Ctenicera pruinina, 84
cultivar, 2, 9, 10, 17, 18, 20
cultural control, 16, 17, 61
Curculionidae, 26, 49, 101
cutworm, 25, 26, 27, 53, 54, 55
Cyclocephala, 83

D
damsel bug, 98
degree days, 4, 42, 65
Delphacidae, 26, 63
desert locust, 91, 92
direct injury, 5
Diuraphis frequens, 44
Diuraphis noxia, 42
Diuraphis tritici, 44
Dolerus spp., 62

E
economic injury level, 12, 14, 15
economic threshold, 14, 15, 41, 43
Elasmopalpus lignosellus, 64
Elateridae, 26, 84, 104, 105
Eleodes extricata, 85
Eleodes hispilabris, 85
Eleodes opacus, 85
Eleodes suturalis, 85
Embaphion muricatum, 85
English grain aphid, 20, 34, 39, 40, 44, 45
entomopathogens, 1, 93
Eriophyes tulipae, 75
European corn borer, 25, 26, 29, 32, 68, 69
European wheat stem sawfly, 80, 81

F
fall armyworm, 46, 47, 49
false wireworm, 85
Faronta diffusa, 47
flea beetle, 55
flower fly, 97
fly-free date, 47
Frankliniella fusca, 73
frit fly, 56
fungicides, 10

G
glume blotch, 10
grasshopper, 16, 24, 50, 56, 57, 58
grass sheathminer, 63, 64
greenbug, 17, 18, 19, 39, 40, 41, 45, 99
green lacewing, 97
ground beetle, 87, 88, 99
ground pearls, 88, 89
growing degree days, 4

H
Hemerobiidae, 97
Hemiptera, 26, 37, 38, 39, 40, 42, 43, 52, 63, 66, 70, 86, 88, 98
Hessian fly, 11, 13, 14, 16, 17, 18, 19, 23, 25, 26, 27, 28, 29, 33, 47, 52, 58, 59, 60, 61, 62, 73, 89, 94, 96
Heteroptera, 26, 52, 66, 70, 86, 98
high plains virus, 21, 75, 77
Hippodamia convergens, 96
Homoptera, 26
Hymenoptera, 26, 62, 73, 75, 80, 82, 94, 95

I
importation, 18
indirect injury, 12
insecticide, 1, 4, 16, 17, 19, 40, 41, 42, 43, 53, 54, 58, 65, 73, 75, 82, 84, 87, 100
integrated pest management, 1
IPM, 1, 12, 14, 15, 19
Irbisia, 66

J
June beetle, 30

K
kernel, 5, 6, 7, 58, 70, 74, 75, 86

L
Labops, 66, 67
lacewing, 97, 98
lady beetle, 96
leafhoppers, 21, 63
leafminers, 64
Leptopterna, 66, 67
lesser cornstalk borer, 64
Limonius spp., 84
Limothrips cerealium, 72
Lopus, 66, 67
lotus borer
Lygaeidae, 26, 52
Lygus, 66, 67

M
Macroglenes penetrans, 75
Macrosteles quadrilineatus, 21, 63
masked chafer, 83
Mayetiola destructor, 58
May or June beetles, 83
Melanoplus spp., 57
Meloidae, 26, 49, 58
Meromyza americana, 79
Metopolophium dirhodum, 20, 44, 45
midge, 16, 25, 26, 29, 33, 73, 74, 75, 89
Miridae, 26, 66
mite, 1, 14, 16, 18, 21, 22, 25, 26, 35, 75, 76, 77, 78, 79, 97
Mormon cricket, 24, 25, 26, 30, 65, 66

N
Nabidae, 98
Nabis spp., 98
natural enemies, 1, 18, 19, 23, 38, 43, 47, 51, 63, 100
no tillage, 110

O
oats, 1, 2, 3, 4, 6, 7, 10, 11, 20, 21, 22, 39, 40, 50, 51, 56, 63, 65, 68, 70, 72, 75, 79, 80, 86, 90
oat blue dwarf virus, 63
Oligonychus pratensis, 78
Orthocephalus, 66, 67

Orthoptera, 26, 57, 65, 91
Ostrinia nubilalis, 68
Oulema melanopus, 50

P
Pachynematus, 62
pale western cutworm, 54
Pandora neoaphidis, 93, 94
Papaipema nebris, 69
parasitoids, 1, 18, 19, 23, 38, 41, 43, 51, 54, 55, 58, 60, 63, 65, 71, 73, 80, 82, 83, 94, 95, 96, 100
pathogens, 1, 12, 18, 19, 21, 22, 23, 37, 41, 43, 54, 55, 58, 65, 80, 93, 94
Pentatomidae, 26, 70, 86
Penthaleus major, 79
pest management, 1, 12, 14, 15, 17, 19, 24, 93, 100
Petrobia latens, 77
Phyllophaga, 83, 102
planthopper, 63
plant bugs, 66, 67, 93, 98
plant injury, 12, 13, 16
plant resistance, 14, 17, 89
predators, 1, 18, 38, 41, 43, 46, 54, 55, 58, 65, 71, 95, 97, 98, 99, 100
Pseudaletia unipuncta, 46, 102
Pyralidae, 26, 64

R
radish, 22, 23
redlegged grasshopper, 57
Reduviidae, 98
Rhopalosiphum maidis, 38
Rhopalosiphum padi, 20, 37
Rhopalosiphum rufiabdominalis, 20, 44
rice hoja blanca virus, 63
rice root aphid, 20, 44, 45
roots, 4, 5, 6, 12, 22, 45, 49, 52, 59, 83, 84, 86, 87
rose grain aphid, 44
Russian wheat aphid, 16, 17, 18, 19, 27, 28, 34, 42, 43, 44, 93, 94, 98
rye, 1, 2, 4, 7, 10, 11, 17, 18, 20, 21, 22, 39, 51, 59, 65, 67, 72, 73, 86

S
sawflies, 32, 62, 63, 82
seedcorn maggot, 67, 90
sevenspotted lady beetle, 96
sheathminer, 25, 26, 63, 64
Sipha elegans, 44, 45
Sipha flava, 43

Sitobion avenae, 20, 39
Sitodiplosis mosellana, 73
spider, 97, 99
Spodoptera frugiperda, 46
Spodoptera ornithogalli, 48
spring wheat, 6, 7, 11, 14, 17, 39, 43, 45, 46, 49, 60, 61, 73, 74, 75, 80, 82, 84, 87
stalk borer, 32, 69, 70
Staphylinidae, 98
stem weevil
Stenotus, 66, 67
stink bug, 70, 71, 72, 86
sunn pest, 86, 87
Syrphidae, 97
syrphid fly, 40, 41

T
Tachinidae, 54, 55
Tenthredinidae, 26, 62
Tetranychidae, 26, 77, 78
Thripidae, 26, 72
thrips, 22, 23, 28, 72, 73, 98
tillage, 9, 10, 16, 19, 22, 49, 52, 58, 61, 68, 73, 82, 100
Trachelus tabidus, 80
Trigonotylus, 66, 67
triticale, 1, 2, 4, 7, 10, 17, 21, 43, 59

V
vector, 20, 21, 38, 40, 41, 63, 75
virulence, 17, 61, 93
viruliferous, 23
virus, 11, 18, 20, 21, 22, 38, 39, 40, 63, 65, 75, 76, 77

W
weeds, 1, 10, 11, 15, 16, 22, 23, 58, 69, 76, 80, 86, 87
weevils, 49
western wheat aphid, 44
wheat curl mite, 16, 18, 21, 22, 75, 76, 77
wheat ground beetle, 87, 88
wheat head armyworm, 48
wheat jointworm, 23, 73
wheat stem maggot, 80, 82
wheat stem sawfly, 14, 16, 17, 18, 23, 80, 81, 94
wheat strawworm, 23, 83
wheat streak mosaic virus, 18, 21, 77
white grub, 84
winter grain mite, 79

winter wheat, 2, 6, 7, 9, 10, 11, 13, 14, 16, 20, 23, 40, 43, 60, 61, 66, 68, 75, 76, 78, 80, 82
wireworm, 27, 30, 36, 84, 85

X
Xanthomonas campestris, 22, 40

Y
yellowstriped armyworm, 48, 49
yellow sugarcane aphid, 43, 44, 45

About the Handbook Series

Handbook of Small Grain Insects represents the seventh in a series of handbooks developed to serve IPM practitioners and educators requiring a comprehensive and thorough source of information in a particular area or commodity. The Entomological Society of America continues its long standing service, not only to its members through the publication of original research, but also through providing useful reference materials for growers, extension agents, consultants, agronomists, veterinarians, master gardeners, pest control operators, researchers, and students.

This handbook series provides pest information, including life cycle, host plant damage, and pest distribution, and contains excellent color photographs, line drawings, and pest distribution maps that aid in identification. Each handbook in the series focuses on a particular commodity, a specific habitat, or particular group of pests. Written with the non-specialist in mind, each handbook thoroughly covers each pest in detail. The authors of each pest section are authorities from throughout the United States and present the latest in research information and knowledge.

The Entomological Society of America's commitment to the highest quality handbook series should be evident as one reads *Handbook of Small Grain Insects*. This handbook contains some of the most comprehensive and up-to-date information on managing small grain insects throughout the United States and Canada. The first three sections of the handbook provide information about small grains and their production, principles and practices of small grain insect management, and an overview of the pest injury to small grains by insects, weeds, and plant pathogens, and includes identification keys of insect pests. The remainder of the book is devoted to discussions of insects and mite pests of small grains and to beneficial organisms. Authors are identified at the end of each section, and a list of references for additional information is provided for most sections. A glossary provides definitions for unfamiliar words and a section is provided on sources of local information throughout the United States and Canada.

This welcome addition to the series is a result of the combined efforts of many individuals including the handbook editors, David Buntin, Keith Pike, Michael Weiss, and James Webster, as well as 51 individual section authors. These individuals graciously devoted their time, energy, and expertise to the development of this comprehensive handbook. Special thanks are in order for the time and effort expended by the members of the handbook editorial committee: John All, Fred Baxendale, Jack DeAngelis, Mike Gray, David Shetlar, Donald Booth, Marlin Rice, Leon Higley, Daniel Potter, and Michael Villani.

Two other new handbooks were published recently—*Handbook of Forage and Rangeland Insects* and *Use and Management of Insecticides, Acaricides, and Transgenic Crops*. In addition, watch for a new handbook covering insect pests of pets and small animals.